Science
Activity Workbook

Grade 4

Cover: Burnett/Palmer/Mira; orange tube coral border—Ed Robinson/Pacific Stock, Inc.
Title page: Ed Robinson/Pacific Stock, Inc.

Published by Macmillan/McGraw-Hill, of McGraw-Hill Education, a division of The McGraw-Hill Companies, Inc., Two Penn Plaza, New York, New York 10121. Copyright © by Macmillan/McGraw-Hill. All rights reserved. The contents, or parts thereof, may be reproduced in print form for non-profit educational use with Macmillan/McGraw-Hill Science, provided such reproductions bear copyright notice, but may not be reproduced in any form for any other purpose without the prior written consent of The McGraw-Hill Companies, Inc., including, but not limited to, network storage or transmission, or broadcast for distance learning.

Printed in the United States of America

5 6 7 8 9 024 09 08 07

Table of Contents

Unit A   **The World of Living Things**

Chapter 1—From Cells to Ecosystems 1

Chapter 2—Plants 29

Unit B   **Animals as Living Things**

Chapter 3—Describing Animals 39

Chapter 4—Life Processes 54

Unit C   **Earth and Beyond**

Chapter 5—Earth's History 69

Chapter 6—Earth's Surface and Interior 80

Chapter 7—Sun, Moon, and Planets 93

Unit D   **Water and Weather**

Chapter 8—Earth's Water 103

Chapter 9—Earth's Weather 130

Unit E   **Matter**

Chapter 10—Properties of Matter 138

Chapter 11—Changes in Matter 150

Unit F   **Energy**

Chapter 12—Forms of Energy 165

Chapter 13—Electricity and Magnetism 188

©Macmillan/McGraw-Hill

Name_____ Date_____

What Are Living Things Made Of?

Hypothesize Sometimes it is hard to tell if an object is living or nonliving. However, living things have certain parts in common. What might they be? Write a **Hypothesis:**

_____

_____

Materials
- onion plant
- prepared slides of onion skin and leaf
- hand lens
- microscope

Procedure

1. **Observe** On a separate sheet of paper, draw the whole onion plant. Label its parts. Write down how each part might help the plant live.

2. **Observe** Ask your teacher to cut the plant lengthwise. On another sheet of paper, draw and label what you see.

3. **Observe** Look at a small section of onion skin and a thin piece of a leaf with the hand lens. On another sheet of paper, draw what you see.

4. Use the microscope to look at the onion skin and the leaf section. Use high and low power. On another sheet of paper, draw what you see.

Drawing Conclusions

1. **Communicate** What did you see when you examined the onion skin and leaf with the hand lens and the microscope? Make a table or chart.

Unit A · The World of Living Things   Use with textbook page A5

Name_____ Date_____

2. **Infer** What do the parts of the onion plant seem to be made of?
_____

3. FURTHER INQUIRY **Predict** Do you think you would see similar structures if you observed a part of the root? How could you test your prediction? Try it and report your results.
_____
_____

Inquiry

Think of your own questions about parts of plants that you might like to test. Would other plants have structures similar to the ones you saw in the onion plant?

My Question Is:
_____
_____

How I Can Test It:
_____
_____

My Results Are:
_____
_____

Name_____ **Date**_____

Alternative Explore
Lesson 1

Speaking Cells

Procedure

1. Use the microscope to look at the slide of onion skin.

2. Describe for your partner what you see. Have your partner draw the cells based on your description, without looking through the microscope.

3. While your partner describes the slide of the onion leaf, draw the cells based on that description. Do not look through the microscope.

4. After you have both made your drawings, use the microscope to look at the slides and compare what you see to your drawing.

Materials

• prepared slides of onion skin and leaf

• microscope

Drawing Conclusions

1. Which cell did you draw?

_____

2. Did your drawing look like the cells that were described to you? Explain.

_____

_____

©Macmillan/McGraw-Hill

Unit A · The World of Living Things   Use with TE textbook page A5   **3**

Name_____ Date_____

Inquiry Skill Builder
Lesson 1

Make a Model

Plant and Animal Cells

Use a microscope to look at plant and animal cells. The plant cells are from *Elodea*, a freshwater plant. The animal cells are cheek cells from inside a person's mouth. Look for the cell parts you learned about. Then make models to show differences between plant and animal cells. Models are three-dimensional copies of real objects. They help you see how things are put together.

Procedure

1. **Observe** Place the slide of *Elodea* on the microscope stage. Focus through the top layers of cells using low power. Observe one cell carefully.

2. **Infer** Record your observations by making a labeled drawing. Identify the names of what you see based on what you have learned about cells.

Materials

- prepared slide of *Elodea* leaf
- prepared slide of human cheek cells
- microscope
- art materials and common objects for cell models
- tied plastic sandwich bags filled with light-colored gelatin (optional)

© Macmillan/McGraw-Hill

4   Unit A · The World of Living Things   Use with textbook page A11

Name _____ **Date** _____

Inquiry Skill Builder
Lesson 1

3. Repeat steps 1 and 2 with the slide of human cheek cells.

4. **Make a Model** Decide on how you will represent each kind of cell to show its shape and parts. List the parts. Select objects you might use to show each part, such as a clear plastic sandwich bag, lima beans, or marbles. To start, you might use gelatin as the cytoplasm. Build your models.

Drawing Conclusions

1. How are your models different from the real cells you observed? How are the models like the cells you observed?

   _____

   _____

   _____

2. How might you combine parts from all the models you looked at to make the best cell model of all?

   _____

   _____

©Macmillan/McGraw-Hill

Unit A · The World of Living Things   Use with textbook page A11   5

Name _____ Date _____

How Are Organisms Classified?

Hypothesize What characteristics do you think scientists use to classify living things? Write a **Hypothesis**:

_____
_____
_____

Materials
- reference books

Procedure

1. Choose eight very different organisms that you would like to classify and learn more about. You may choose the ones you see on the third page of this Explore Activity. Record their names.

   _____
   _____

2. What would you like to know about your organisms? Where would you look to find the information? Design a table to record the information.

3. **Classify** Try to place all of the organisms into groups. What characteristics did you use to help you make your choices?

   _____
   _____

6   Unit A · The World of Living Things   Use with textbook page A19

Name_____ Date_____

Drawing Conclusions

1. **Interpreting Data** How many groups were formed? What were the major characteristics of the organisms in each group?
   _____
   _____

2. What organisms were placed in each group?
   _____
   _____

3. **Communicate** Make a list of the characteristics of the organisms in each group.
   _____
   _____

4. FURTHER INQUIRY **Classify** Use your table to explain how you would classify an organism that you had never seen before.
   _____
   _____

Inquiry

Think of your own questions concerning classification that you would like to research.

My Question Is:
_____

How I Can Test It:
_____

My Results Are:
_____
_____

Unit A · The World of Living Things — Use with textbook page A19

Name_____ Date_____

Explore Activity
Lesson 2

_____   _____

_____   _____   _____

_____   _____

8 | Unit A · The World of Living Things | Use with textbook page A19

Name_____ Date_____

Alternative Explore
Lesson 2

Classify Seeds

Procedure

Materials
• A mixture of 10 different kinds of seeds

1. Your teacher will give you and your partner a mixture of seeds. Examine the seeds. Look for ways they are alike and different.

2. Create a classification system for the seeds. Begin by dividing the seeds into two groups based on a single characteristic.

3. Use the two larger blank boxes in the diagram below to list the seeds in the two groups.

4. Break each of the two groups into two sub-groups based on another characteristic.

5. Use the four smaller boxes in the chart to list the characteristics of the seeds in each of the smaller groups.

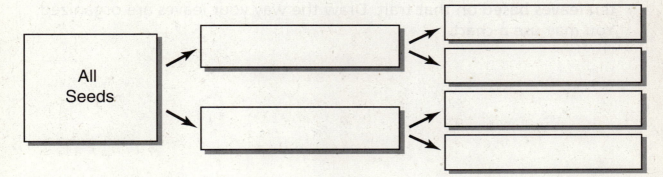

Drawing Conclusions

1. What characteristic did you use in step 2?
_____

2. What characteristics did you use in step 4?
_____

3. How does classifying the seeds make it easier to study them?
_____
_____

Unit A · The World of Living Things   Use with TE textbook page A19

Name_____ **Date**_____

Inquiry Skill Builder
Lesson 2

Classify

Leaves

When you organize toys or living things into groups, you are classifying. When you classify, you organize things into smaller groups based on their traits. This skill is important not only in science. People classify things every day. Classifying helps make things easier to study and understand. To practice this skill, you will classify leaves according to different traits.

Materials

- 10 leaves or leaf pictures shown on the third page of this Skill Builder
- ruler
- reference books

Procedure

1. **Observe** Spread out the leaves (or leaf pictures). Observe the traits they share, such as size, color, shape, and so on. Record the traits.

   _____

2. **Classify** Choose one trait, such as color, that you recorded. Organize all ten leaves based on that trait. Draw the way your leaves are organized. You may use a chart.

10   Unit A · The World of Living Things   Use with textbook page A24

©Macmillan/McGraw - Hill

Name _____ **Date** _____

Inquiry Skill Builder
Lesson 2

3. Follow the same procedure for two other traits you recorded.

Drawing Conclusions

1. In how many different ways were you able to classify the leaves?

   _____

2. How did your classification system differ from other students' systems? In what ways were they similar?

   _____

   _____

3. **Classify** Group the leaves by their function. How many groups did you come up with? Explain the function of leaves.

   _____

   _____

4. **Infer** How do you think using a worldwide classification system might help scientists identify and understand organisms?

   _____

   _____

©Macmillan/McGraw-Hill

Unit A · The World of Living Things   Use with textbook page A24   11

Name_____ Date_____

Maple

Oak

Walnut

Cherry
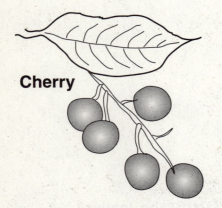

Mulberry

Juniper

Pine
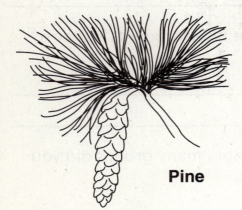

Redwood
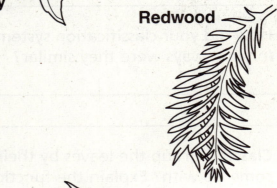

Apple

Sycamore
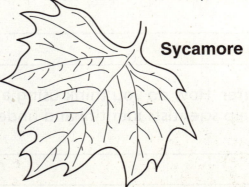

12 | Unit A · The World of Living Things | Use with textbook page A24

Name _____ Date _____

Explore Activity
Lesson 3

How Can We Use Skeletons to Compare Organisms?

Materials
- ruler, pencil, and paper, or computer with charting program

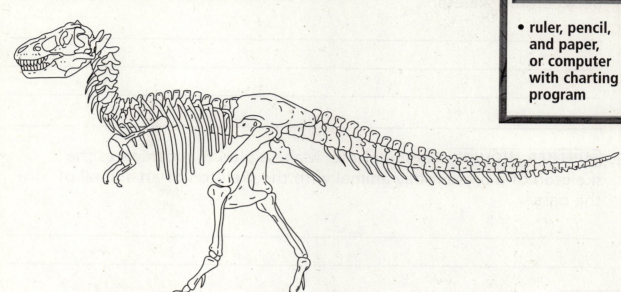

Hypothesize Does this dinosaur skeleton remind you of any animals alive today? What might we learn from comparing the bones of animals from the past with those of animals living today? Write a **Hypothesis:**

_____

_____

Procedure

1. **Observe** Compare the picture of the dinosaur skeleton with the bird and reptile skeletons on the third page of this Explore Activity.

2. **Communicate** Make a chart that lists the similarities and differences. Use the computer if you like.

Unit A · The World of Living Things — Use with textbook page A29 — 13

Name_____ Date_____

Drawing Conclusions

1. Write a paragraph about the similarities and differences you noticed between the skeletons.

   _____
   _____
   _____
   _____

2. **FURTHER INQUIRY** Communicate Make a plan that compares the skeleton of a present-day animal with the skeleton of an animal of the past.

   _____
   _____

Inquiry

Think of your own questions that you might research about how animals have changed over time. Would you like to compare the environments of an ancient animal and a present-day animal?

My Question Is:

_____

How I Can Test It:

_____

My Results Are:

_____
_____
_____

14 | Unit A · The World of Living Things | Use with textbook page A29

Name_____ Date_____

Explore Activity
Lesson 3

This is a skeleton of a bird.

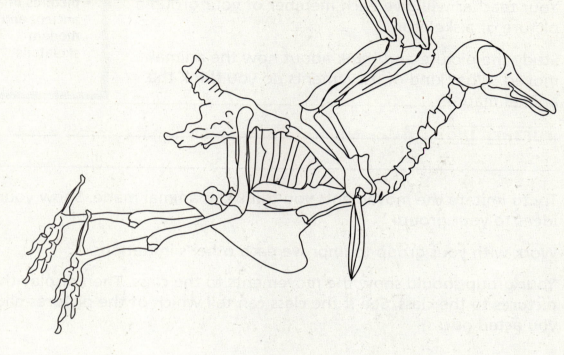

This is a skeleton of a reptile.

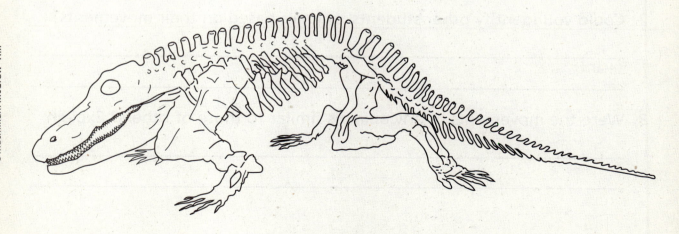

Unit A · The World of Living Things   Use with textbook page A29

Name_____ Date_____

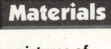
Lesson 3

Compare Skeletons

Procedure

Materials
- pictures of ancient and modern skeletons

1. Your teacher will give each member of your group a picture of a skeleton.

2. Study the picture and think about how the animal moved. What kind of movements do you think the animal made?
   _____
   _____

3. Try to imitate the movements you think the animal made. Show your ideas to your group.

4. Work with your group to improve each other's imitations.

5. Your group should show the movements to the class. Then display the pictures to the class. See if the class can tell which of the pictures each of you acted out.

Drawing Conclusions

1. Could the class identify your animal based on your movements?
   _____
   _____

2. Could you identify other students' animals based on their movements?
   _____
   _____

3. Were the movements of any animals similar to those of others? Explain.
   _____
   _____

Name_____ Date_____

QUICK LAB

FOR SCHOOL OR HOME

Lesson 3

Older and Younger

Hypothesize Relative dating is placing things in order from oldest to youngest. What do you look for to help you decide which is older and younger? Write a Hypothesis:

_____

_____

_____

Materials

• 4 books

• 2 pieces of paper

• scissors

• pen or pencil

Procedure

1. **Observe** Cut a piece of paper into four pieces. Draw a "fossil" on each piece. Have a partner place one fossil inside the front cover of each book. Stack the books.

2. **Interpret Data** Find the fossils and arrange them in order from "oldest" to "youngest." Record any observations you make.

_____

_____

Drawing Conclusions

3. What did the books represent?

_____

4. On a separate piece of paper, draw the fossil that was the oldest. Then draw the fossil that was the youngest. Explain what evidence helped you decide.

_____

_____

_____

©Macmillan/McGraw-Hill

Unit A · The World of Living Things   Use with textbook page A31   17

Name_____ Date_____

5. **Going Further** Carbon dating is a technique used to approximate how long ago an organism died. Research and explain carbon dating.

My Hypothesis Is:

_____

_____

My Experiment Is:

_____

_____

_____

My Results Are:

_____

_____

_____

_____

Unit A · The World of Living Things   Use with textbook page A31

Name_____ Date_____

Explore Activity
Lesson 4

How Do Living and Nonliving Things Interact?

Hypothesize The key to an organism's survival is how it interacts with other living and nonliving things. What might these interactions be? Write a **Hypothesis:**

_____

_____

_____

_____

_____

Materials

• prepared terrarium container

• small plants and animals

• plastic spoon

• ruler

• water mister

• grass seeds, rocks, twigs, sticks, bark, dried grass

Procedure

BE CAREFUL! Handle animals and plants carefully.

1. **Make a Model** Landscape your terrarium. Put taller plants in the back. Spread grass seed and any rocks, twigs, or other things you like.

2. If you add small animals, such as earthworms, sow bugs, and snails, add a water dish.

3. **Measure** Make a data table on a separate piece of paper. Record the height of each plant. Measure the plants in two weeks, and record the data. Make a bar graph.

4. Place the terrarium in a lighted area. Avoid direct sunlight.

5. **Communicate** Draw a diagram of your terrarium. Draw arrows to show how the organisms depend on one another.

©Macmillan/McGraw-Hill

Unit A · The World of Living Things   **Use with textbook page A39**   19

Name_____ Date_____

Explore Activity — Lesson 4

Drawing Conclusions

1. **Classify** What are the living and nonliving things in the terrarium?
   _____

2. **Infer** Why should the terrarium not be placed in direct sunlight?
   _____
   _____
   _____

3. **FURTHER INQUIRY** **Observe** What effect might a new organism have on your terrarium ecosystem? Add an organism, observe, and report your results.
   _____
   _____

Inquiry

Think of how you might change the terrarium ecosystem. Would you like to add or remove a living or nonliving thing?

My Question Is:

_____
_____

How I Can Test It:

_____

My Results Are:

_____
_____

20 Unit A · The World of Living Things   Use with textbook page A39

Name_____ **Date**_____

Alternative
Explore
Lesson 4

Terrarium Ecosystems

Procedure

1. Choose an environment to model in a terrarium.

2. Look up this environment in books to find out what the soil is like and what the climate is like.

3. Choose soil and plants for your terrarium.

4. Assemble your terrarium. Add some water to the soil.

5. Cover the terrarium and place it in a lighted area, but not in direct sunlight.

6. Observe your terrarium for several days. If you think your environment is too dry, add water. If you think it is too wet, leave the cover off for a few hours to allow some water to evaporate.

Materials

- terrarium container
- sand
- loam
- clay
- rocks
- small plants
- plastic spoon
- water

Drawing Conclusions

1. What environment did you choose?

_____

2. What is the climate like in this environment?

_____

3. What plants grow in this environment?

_____

4. Did you have to adjust the amount of water in your terrarium? Why or why not?

_____

_____

©Macmillan/McGraw-Hill

Unit A · The World of Living Things   Use with TE textbook page A39   **21**

Name_____ **Date**_____

QUICK LAB
FOR SCHOOL OR HOME
Lesson 4

Sunlight and Plants

Hypothesize What kinds of habitat does a plant prefer?
Form a hypothesis. Use two plants to find out. Write a
Hypothesis:

_____

_____

Materials
- two identical plants
- two labels
- paper bag
- water

Procedure

1. Your teacher will give you two identical plants. Label one plant *No Light.* Place it on a table away from a window. Water the plant. Then cover it with a paper bag.

2. Label the other plant *Light.* Place it on a sunny windowsill or near a sunny window. Water the plant.

3. **Observe** Every other day for two weeks, observe the plants. Check to make sure the plants have moist soil. Record changes in the color and height of each plant.

Drawing Conclusions

4. Which plant looked the healthiest after two weeks?

   _____

5. What do your observations tell you about the needs of plants?

   _____

   _____

©Macmillan/McGraw-Hill

22   Unit A · The World of Living Things   Use with textbook page A41

Name_____ Date_____

QUICK LAB
FOR SCHOOL OR HOME
Lesson 4

6. **Going Further** This experiment used growing conditions with more and less light. What would happen if you used warmer or cooler conditions, or wetter or drier conditions? Describe and perform an experiment to test a specified environment.

My Hypothesis Is:

_____

_____

My Experiment Is:

_____

_____

_____

_____

My Results Are:

_____

_____

©Macmillan/McGraw-Hill

Unit A · The World of Living Things   Use with textbook page A41   23

Name_____ Date_____

QUICK LAB FOR SCHOOL OR HOME
Lesson 4

Observe a Decomposer

Hypothesize What do you think molds need to grow?

Write a **Hypothesis**:
_____

Procedure

BE CAREFUL! Do not open the bags after you seal them.

1. Moisten four food samples. Place each in a labeled plastic bag. Put a piece of cardboard in a bag.
2. Seal the bags, and place them in a warm, dark place.
3. Record your daily observations.

Materials
- 5 sealable plastic bags
- warm water
- 4 food samples
- hand lens
- piece of cardboard
- marking pen

Observations					
Day	Card-board				
1					
2					
3					
4					
5					

24 Unit A · The World of Living Things   Use with textbook page A49

Name _____ Date _____

Drawing Conclusions

4. **Observe** On which samples did mold grow? How did it change the foods?

   _____
   _____
   _____

5. **Infer** Will molds grow on any type of material? Explain how the cardboard helped you answer this question.

   _____
   _____
   _____

6. **Going Further** Several methods are used to deter the growth of molds on food and extend the time that foods are safe for consumption. Refrigerating, or cooling food, slows the growth of molds. Pickling and salting foods create an environment that molds cannot survive. Test these methods of food preservation.

   My Hypothesis Is:
   _____

   My Experiment Is:
   _____
   _____
   _____
   _____
   _____

   My Results Are:
   _____
   _____

Unit A · The World of Living Things   Use with textbook page A49

Name_____ Date_____

What Causes Ecosystems to Change?

Hypothesize How will a different amount of moisture affect the plants in a terrarium? Write a **Hypothesis:**
_____
_____

Materials
- 2 terrariums (from Lesson 4)
- water mister

Procedure

1. Label one terrarium M, for moist. Label the other D, for dry. Spray only terrarium M with water. Place both terrariums in a well-lighted area, but avoid direct sunlight.

2. **Observe** Compare the terrariums every day for several days. What differences do you see?

3. **Observe** Spray terrarium D until it is moist. Wait a few days before comparing the terrariums.

4. **Communicate** Make a table to record what you observed in steps 2 and 3.

Drawing Conclusions

1. What happened to the organisms in each terrarium in step 2?
_____
_____

2. What happened to the organisms in each terrarium in step 3?
_____
_____

26   Unit A · The World of Living Things   Use with textbook page A55

Name_____ Date_____

3. **Infer** What caused the changes in the terrariums?
   _____
   _____

4. **Infer** What happened to terrarium D in step 3? What does that tell you about ecosystems?
   _____
   _____

5. FURTHER INQUIRY **Experiment** Observe the terrariums for the rest of the year. Periodically bring them outside. What effects do weather and seasons have on living things in the ecosystem?

Inquiry

Think of your own questions that you might like to test. Can plants get enough moisture from water that evaporates into the air from a bowl of water placed in the terrarium?

My Question Is:

_____
_____

How I Can Test It:

_____
_____

My Results Are:

_____
_____
_____
_____

Unit A · The World of Living Things   Use with textbook page A55

Name_____ Date_____

Alternative
Explore
Lesson 5

Changing the Amount of Sunlight

Procedure

1. Place one of the terrariums in a sunny place.

2. Place the second terrarium in a dark place.

3. Observe the plants in the terrariums daily.

4. After a week or so goes by, compare the appearances of the plants in the two terrariums. Record your comparisons.

_____

_____

_____

Materials

• 2 terrariums from Lesson 4

Drawing Conclusions

1. How does sunlight affect plants in a terrarium?

_____

_____

2. How does a lack of sunlight affect plants in a terrarium?

_____

_____

_____

3. How might the plants be affected if you kept the terrariums where they are for another week?

_____

_____

_____

28   **Unit A · The World of Living Things**   **Use with TE textbook page A55**

Name_____ Date_____

Explore Activity
Lesson 6

Why Does a Plant Need Leaves?

Hypothesize What will happen to a plant over time if its leaves are removed? Write a **Hypothesis:**

_____

_____

Materials

- 2 potted plants of the same kind with leaves
- scissors
- metric ruler
- index card or paper
- tap water

Procedure

BE CAREFUL! Handle scissors carefully.

1. Label plants A and B. Use the scissors to cut all the leaves from plant B.

2. Water the soil in both plants. Place the plants in a well-lighted area, but not in direct sunlight. Do not move the plants during the experiment.

3. **Observe** Examine the plants each day for ten days. Measure any growth, and note any other changes. Record your observations in the table below.

Observations

DAY	PLANT A	PLANT B
1		
2		
3		
4		
5		
6		
7		
8		
9		
10		

Unit A · The World of Living Things   Use with textbook page A69

Name_____ Date_____

Drawing Conclusions

1. **Interpret Data** What happened to plant A? To plant B?
   _____
   _____

2. **Form a Hypothesis** Why does a plant need leaves?
   _____
   _____

3. FURTHER INQUIRY **Predict** What would happen if you kept plant A in a dark place? Try it and report your results.
   _____

Inquiry

Think of your own questions that you might like to test. Can a leaf live without the plant?

My Question Is:
_____

How I Can Test It:
_____
_____

My Results Are:
_____
_____

30 unit A · The World of Living Things   Use with textbook page A69

Name _____ **Date** _____

Alternative Explore
Lesson 6

Comparing Plant Leaves

Procedure

1. Compare the different plant leaves.

2. On a sheet of paper, draw each leaf. Record your observations.

_____

_____

_____

_____

Materials

- plants with different kinds of leaves (or pictures of different leaves)

Drawing Conclusions

1. What is the same about the leaves?

_____

_____

_____

2. What is different about the leaves?

_____

_____

_____

_____

©Macmillan/McGraw-Hill

Unit A · The World of Living Things   Use with TE textbook page A69   31

Name_____ Date_____

Inquiry Skill Builder
Lesson 6

Predict

Photosynthesis Versus Respiration

Photosynthesis gives off oxygen; respiration gives off carbon dioxide. You can observe this for yourself by placing a leaf in water. The leaf will "breathe," releasing tiny air bubbles that you can see. Which do you predict will produce more air bubbles—a leaf placed in sunlight or darkness?

Materials

- 2 jars
- 2 leaves from the same plant
- water

Procedure

1. Fill each jar with fresh water. Place a leaf from the same plant in each jar.

2. Place one jar in a bright, sunny window. Place the other jar in the dark.

3. **Predict** Write down what you think will happen. On which leaf will bubbles form first?

_____

4. **Observe** Check each leaf in five minutes, keeping a written record of what you observe.

_____

_____

Drawing Conclusions

1. How well did your results agree with your prediction?

_____

_____

2. How do your results for the leaf in sunlight compare with your results for the leaf kept in the dark?

_____

_____

32   Unit A · The World of Living Things   Use with textbook page A75

©Macmillan/McGraw-Hill

Name_____ **Date**_____

Inquiry Skill Builder
Lesson 6

3. **Predict** What do you think will happen if you continue the experiment for one hour? Test your prediction.

_____

_____

4. **Interpret Data** What do the results tell you about photosynthesis and respiration?

_____

_____

_____

Unit A · The World of Living Things   Use with textbook page A75   33

Name_____ Date_____

Explore Activity
Lesson 7

How Does a Seed Grow?

Hypothesize What will happen if you give plant seeds light and moisture? Think about what will happen first, then what changes you might see over ten days. Write a **Hypothesis:**

_____

_____

Materials

- 2 soaked pinto bean seeds
- paper towels
- clear plastic cup
- marking pen
- water

Procedure

1. Fold a paper towel to match the height of the plastic cup. Line the cup with the paper towel.

2. Crumple another paper towel into a ball. Place it in the cup to hold the lining in place.

3. On either side of the cup, place a seed between the lining and the cup, about $\frac{3}{4}$ of the way up.

4. On either side of the outside of the cup, write Seed A near one seed and Seed B near the other.

5. **Observe** Place the cup in a sunny spot. Keep the lining moist. Observe the seeds for seven to ten days. Record your observations in the table.

Plant Observations

DAY	SEED A	SEED B
1		
2		
3		
4		
5		
6		
7		
8		
9		
10		

Unit A · The World of Living Things

Name_____ Date_____

Drawing Conclusions

1. **Communicate** Describe the plant parts that you observed.
   _____
   _____

2. **Interpret Data** Analyze your chart for patterns and changes. What does a seed need to grow into a plant?
   _____

3. FURTHER INQUIRY **Use Variables** Try the experiment without adding water to the cup. Report your results.
   _____

Inquiry

Think of your own questions that you might like to test. How would a sprouting seed be affected if it were placed in a very cold or a very warm place?

My Question Is:
_____
_____

How I Can Test It:
_____
_____
_____

My Results Are:
_____
_____
_____

Unit A · The World of Living Things   Use with textbook page A81

Name_____ **Date**_____

Alternative Explore
Lesson 7

First Things First

Procedure

1. Describe what is happening in each picture.

_____

_____

_____

2. Tape or staple the pictures of the peach life cycle in the correct order below.

Materials

- photocopies of the illustration of the peach life cycle, with the four stages cut apart

Drawing Conclusions

How does the peach change during its life cycle?

_____

_____

©Macmillan/McGraw-Hill

36   Unit A · The World of Living Things   Use with TE textbook page A81

Name_____ Date _____

QUICK LAB
FOR SCHOOL OR HOME
Lesson 7

The Structure of a Seed

Hypothesize What parts are inside of a seed?

Write a **Hypothesis:**

_____

_____

Materials

- beans that have been soaked
- paper towel
- hand lens

Procedure

1. Place water-soaked bean seeds on a paper towel. The beans should have been soaked in water overnight.

2. **Observe** Carefully pull apart the two halves of the seed. Examine each half with a hand lens.

3. **Communicate** Draw a picture of the seed parts you observed.

©Macmillan/McGraw-Hill

Unit A · The World of Living Things   Use with textbook page A83   37

Name	Date

QUICK LAB

FOR SCHOOL OR HOME

Lesson 7

Drawing Conclusions

4. Which part of the seed is the seed coat? Which part is the source of the root, stem, and leaves? Label the parts.

_____

_____

5. **Infer** Around the seed sketch, explain where the seed stores its food.

_____

6. **Going Further** Find out how different amounts of light affect the way a seed develops. Soak two beans and place each one in a resealable sandwich bag along with a damp paper towel. Write and conduct an experiment.

My Hypothesis Is:

_____

_____

My Experiment Is:

_____

_____

My Results Are:

_____

_____

©Macmillan/McGraw-Hill

38 **Unit A · The World of Living Things**   **Use with textbook page A83**

Name_____ Date_____

Explore Activity
Lesson 1

What Are Some Animal Characteristics?

Hypothesize Animals come in many sizes and shapes, yet they all have certain characteristics. For example, what are the main characteristics of a fish and snail?

Write a **Hypothesis:**

_____

_____

Materials

- clear container with aquarium water
- water snail
- goldfish or guppy
- fish food
- ruler

Procedure **BE CAREFUL!** Handle animals with care.

1. Obtain a container with a fish and a snail in it.

2. **Observe** Record the shape and approximate size of both animals. Describe how each animal moves and any other observations that you make.

_____

_____

3. Add a few flakes of fish food to the beaker. What do the animals do? Record your observations. _____

_____

4. What does the fish eat? The snail?

_____

Drawing Conclusions

1. **Observe** What body parts does each animal have? How do they use these parts? _____

_____

_____

_____

©Macmillan/McGraw-Hill

Unit B · Animals as Living Things   **Use with textbook page B5**   39

Name_____ Date_____

2. Compare how the fish and the snail move. Is movement an advantage for the animals? Explain.

_____
_____

3. **Infer** Do you think the fish and the snail are made of one cell or many cells? Why?

_____

4. **Communicate** What are some characteristics the fish and the snail have? Make a list. Compare your list with other groups' lists. Make a class list.

_____

5. **FURTHER INQUIRY** **Infer** How are the fish and snail able to live in water? Make a plan to find out.

_____
_____
_____

Inquiry

Think of your own questions that you might like to test. Do fish and snails need oxygen to survive, as people do?

My Question Is:

_____

How I Can Test It:

_____

My Results Are:

_____
_____

40 Unit B · Animals as Living Things   Use with textbook page B5

Name_____ Date_____

Alternative Explore
Lesson 1

Observing Animals

Procedure

Materials

- cards showing pictures of vertebrates and invertebrates

1. Obtain one invertebrate card and one vertebrate card.

2. Notice the structure of each animal. Try to tell how the animal moves, gets its food, and eats its food.

   _____

3. What was your vertebrate?

   _____

4. What body parts does the vertebrate have for moving around?

   _____

5. What was your invertebrate?

   _____

6. What body parts does the invertebrate have for moving around?

   _____

   _____

Drawing Conclusions

1. Do you think these animals are made of one cell or many cells? Why?

   _____

2. What parts do the vertebrate and invertebrate have in common?

   _____

3. How does the way the invertebrate and the vertebrate move help them survive?

   _____

Unit B · Animals as Living Things   Use with TE textbook page B5

Name_____ Date_____

Inquiry Skill Builder
Lesson 1

Observe

Animal Symmetry

A scientist's most important job is to *observe*, meaning to look closely. When you observe carefully, you often see things that you didn't know were there. You can practice your observation skills by looking for symmetry in different animals.

Procedure

1. **Observe** Determine whether each animal shown has no symmetry, radial symmetry, or bilateral symmetry.

©Macmillan/McGraw-Hill

42   Unit B · Animals as Living Things   Use with textbook page B10

Name_____ **Date**_____

Inquiry Skill Builder

Lesson 1

2. **Classify** Record your observations in a chart in the space below.

Drawing Conclusions

1. Which animal or animals have radial symmetry? Bilateral symmetry? No symmetry?

_____

_____

2. **Infer** Does an animal with radial symmetry have a front end and a back end? Explain.

_____

_____

3. Which of these animals do you have most in common? Explain.

_____

©Macmillan/McGraw-Hill

Unit B · Animals as Living Things   **Use with textbook page B10**   43

Name_____ Date_____

Explore Activity
Lesson 2

What Are the Characteristics of Invertebrates?

Hypothesize What do you think invertebrates have in common?

Write a **Hypothesis:**

_____

Materials
- living planarian
- living earthworm
- hand lens
- petri dish
- water
- damp paper towel
- toothpick

Procedure BE CAREFUL! Be careful with live animals.

1. **Observe** Place the worm on the damp paper towel. Get a petri dish with a *planarian* (pluh·NAYR·ee·uhn) in it from your teacher. Observe each organism with a hand lens. Record your observations.

_____

2. Gently touch the worm with your finger and the planarian with the toothpick. What do they do? Record your observations.

_____

3. What characteristics of the praying mantis and magnified hydra do you observe? Record your observations.

_____
_____
_____
_____

Hydra

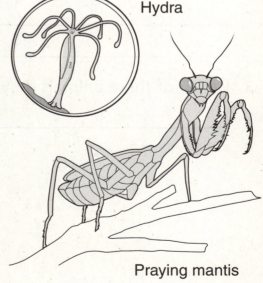
Praying mantis

Unit B · Animals as Living Things   Use with textbook page B15

Name_____ Date_____

Drawing Conclusions

1. **Define Based on Observations** What characteristics do you think invertebrates have? Make a list.
   _____
   _____

2. **Communicate** Compare your list with those of your classmates. Based on your observations, make a class list of invertebrate characteristics.

3. FURTHER INQUIRY **Classify** Think of other organisms that you would classify as invertebrates based on your observations. Make a list. What other questions can you ask to learn more about invertebrates?
   _____
   _____
   _____

Inquiry

Think of your own questions that you might like to test. What size are invertebrates?

My Question Is:
_____
_____

How I Can Test It:
_____
_____

My Results Are:
_____
_____

Unit B · Animals as Living Things   Use with textbook page B15   45

Name_____ Date_____

Compare Phyla

Procedure

1. Your teacher will give your group two or three pictures from one invertebrate phylum.

2. Look at the pictures and discuss the characteristics of each animal.

3. Make a group list of characteristics that all animals in this phylum share.

_____
_____
_____

4. Share your group's list with another group. How are your two phyla similar? How are they different?

_____
_____

Materials
- pictures of invertebrates

Drawing Conclusions

1. Based on your discussion with other groups, what characteristics do invertebrates share?

_____
_____

2. What other animals do you know about that you would classify as invertebrates, based on the list of characteristics you developed in this activity?

_____
_____
_____

46 | Unit B · Animals as Living Things | Use with TE textbook page B15

Name_____ Date_____

Quick Lab
FOR SCHOOL OR HOME
Lesson 2

Classify Invertebrates

Hypothesize What characteristics would you use to classify these invertebrates?

Write a **Hypothesis:**

_____

_____

Procedure

1. **Observe** Use clues in each picture to identify the type of invertebrate.

   _____
   _____
   _____
   _____
   _____
   _____
   _____

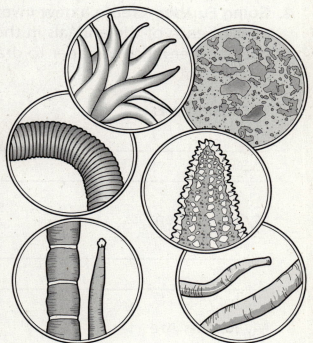

2. **Communicate** Make a table to record how you classified each picture. List key characteristics for each phylum.

Unit B · Animals as Living Things   Use with textbook page B23   47

Name_____ Date_____

Drawing Conclusions

3. Explain how you know the phylum that each animal belongs to.

   _____
   _____
   _____

4. **Going Further** Create a new invertebrate by combining some characteristics of each of the animals in the pictures. Draw the resulting creature. Use another piece of paper to draw the animal. Write and conduct an experiment.

   My Hypothesis Is:

   _____
   _____

   My Experiment Is:

   _____
   _____

   My Results Are:

   _____
   _____
   _____

48   Unit B · Animals as Living Things   Use with textbook page B23

Name_____ Date_____

Explore Activity
Lesson 3

What Are Vertebrates Like?

Hypothesize What characteristics are used to classify vertebrates?

Write a **Hypothesis:**

_____

_____

_____

_____

Procedure: Design Your Own

BE CAREFUL! Handle animals with care.

Observe As you observe each animal, look for answers to these questions. Write down three of your own questions to explore. Record your observations. Try recording sounds or taking photographs to better observe the animals.

Materials

- goldfish
- frog
- chameleon, turtle, or lizard
- parakeet
- hamster, gerbil, or guinea pig
- hand lens
- camera (optional)
- tape recorder (optional)

	Animal A	Animal B	Animal C
a. Where does it live—in water, on land, or both?			
b. What color is it?			
c. What kind of outer covering does it have?			
d. What body parts does it have?			
e. Do you see eyes, ears, nostrils, or other sense organs?			
f. How does it move?			

© Macmillan/McGraw-Hill

Unit B · Animals as Living Things   **Use with textbook page B27**   49

Name_____ Date_____

Drawing Conclusions

1. What major characteristics did you observe in each animal?
   _____

2. What are the main differences between a fish and a frog?
   _____
   _____
   _____

3. What are the major differences between a bird and a hamster?
   _____
   _____
   _____

4. **FURTHER INQUIRY** Classify Which animal in this activity are you most like? Use evidence from the activity to support your explanation.
   _____
   _____
   _____

Inquiry

Think of your own questions that you might like to test. Are some vertebrates more complex than others?

My Question Is:
_____

How I Can Test It:
_____

My Results Are:
_____
_____
_____

50   Unit B · Animals as Living Things   Use with textbook page B27

Name_____ Date_____

Alternative Explore
Lesson 3

Comparing Characteristics

Procedure

1. Your teacher will give your group three pictures from a vertebrate phylum.

2. Look at the pictures and answer the following questions about each animal:

Materials

• pictures of vertebrates

	Picture 1	Picture 1	Picture 1
Where does it live—in water, on land, or both?			
What color is it?			
What kind of outer covering does it have?			
What body parts does it have?			
Do you see eyes, ears, nostrils, or other sense organs?			

3. Share your pictures and lists of characteristics with another group.

Drawing Conclusions

1. What major characteristics did all of your vertebrates share?

_____

2. What differences were there between your phylum and the other group's phylum? _____

_____

3. In what ways are you like the vertebrates in your phylum? In what ways are you different? _____

_____

Unit B · Animals as Living Things   **Use with TE textbook page B27**   **51**

©Macmillan/McGraw-Hill

Name_____ Date_____	**QUICK LAB**
	FOR SCHOOL OR HOME
	Lesson 3

Classify Vertebrates

Hypothesize What characteristics are used to classify these vertebrates?

Write a **Hypothesis:**

_____
_____

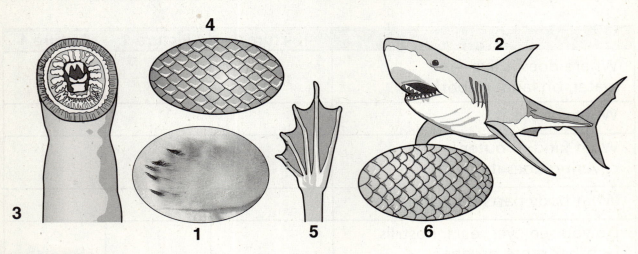

Procedure

1. **Classify** Use the clues in each picture to help you classify each animal.

_____
_____
_____

2. **Communicate** Make and label a table to show how you classified each picture.

52   Unit B · Animals as Living Things   Use with textbook page B35

Name_____ **Date**_____

Drawing Conclusions

3. How do you know which class each animal belongs to?

_____

_____

_____

_____

_____

4. Going Further If you could select a vertebrate characteristic to try for a day, which one would you choose? Why? Write and conduct an experiment.

My Hypothesis Is:

_____

My Experiment Is:

_____

My Results Are:

_____

_____

Unit B · Animals as Living Things   Use with textbook page B35

Name_____ **Date**_____

Explore Activity
Lesson 4

How Does Blood Travel in Fish and Amphibian Hearts?

Hypothesize Which do you think is more complex—a frog's heart or a fish's heart?

Write a **Hypothesis:**

_____

_____

Procedure

1. Label each small cup "atrium." Label each large cup "ventricle."

2. **Make a Model: Fish Heart** Tape the paper circle with one flap over the top of one ventricle. Center the top of an atrium over the flap in the circle. Tape it to the paper.

3. Label one straw "From gills and body." Place it in the hole in the bottom of the atrium. Label another straw "To gills and body." Place it in the hole in the bottom of the ventricle. Draw the model.

4. **Make a Model: Amphibian Heart** Tape the paper circle with two flaps over the top of a ventricle. Center the top of an atrium over each flap. Tape the cups to the paper.

5. Label one straw "From body." Place it in the hole in the bottom of the right cup. Label another straw "From lungs." Place it in the hole in the bottom of the left cup. Label the third straw "To lungs and body." Place it in the hole in the paper between the two small cups. Draw the model on a separate piece of paper.

Materials

- 5 straws

- two 7-oz cups, each with a hole in the bottom

- three $3\frac{1}{2}$-oz cups, each with a hole in the bottom

- 2 paper circles with flaps

- 5 labels

- marking pen

- tape

©Macmillan/McGraw-Hill

54   Unit B · Animals as Living Things   Use with textbook page B45

Name_____ Date_____

Explore Activity
Lesson 4

Drawing Conclusions

FURTHER INQUIRY Observe How do the individual parts of the fish heart interact? The amphibian heart?

_____
_____
_____
_____

Inquiry

Think of your own questions that you might like to test. How do fish and amphibian hearts compare to a human heart?

My Question Is:
_____

How I Can Test It:
_____
_____

My Results Are:
_____
_____

Unit B · Animals as Living Things   Use with textbook page B45   55

Name_____ **Date**_____

Alternative Explore
Lesson 4

Making a Four-Chambered Model

Procedure

1. Use the materials to build a four-chambered heart. Plan how you will build it and make a drawing of your plan.

Materials

- 4 paper cups
- 2 paper circles
- 4 straws
- 4 labels
- tape
- pen

2. Where will you put the straws?

_____

3. Where will you put the paper circles?

_____

_____

Drawing Conclusions

1. What parts of the heart do the cups represent?

_____

2. What structures do the straws represent?

_____

3. What structures do the paper circles represent?

_____

56   Unit B · Animals as Living Things   Use with TE textbook page B45

Name_____ Date_____

QUICK LAB
FOR SCHOOL OR HOME
Lesson 4

Is Bigger Always More?

Hypothesize Can your eyes be fooled?

Write a **Hypothesis:**

_____

A

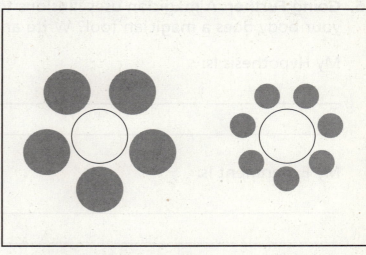

B

Procedure

1. **Observe** Look at drawing A. Record what you see.

   _____
   _____

2. **Observe** Study the center circles in drawing B. Record which is larger.

   _____
   _____

3. **Measure** Measure each center circle. Record which is larger.

   _____
   _____
   _____

Unit B · Animals as Living Things   Use with textbook page B51

Name_____ **Date**_____

QUICK LAB

FOR SCHOOL OR HOME

Lesson 4

Drawing Conclusions

4. Infer Can your eyes fool you? Explain.

_____

_____

5. Going Further A magician uses illusions to perform tricks. Which parts of your body does a magician fool? Write and conduct an experiment.

My Hypothesis Is:

_____

_____

My Experiment Is:

_____

_____

My Results Are:

_____

©Macmillan/McGraw-Hill

58   **Unit B · Animals as Living Things**   **Use with textbook page B51**

Name_____ Date_____

Explore Activity
Lesson 5

How Do Mealworms Change As They Grow?

Hypothesize Do you know of young animals that look very different from their parents? How do you think they change as they grow older?

Write a **Hypothesis:**

_____

_____

Materials

- jars containing food and mealworms in different stages of development

- 3 hand lenses

- 3 rulers

Procedure: Design Your Own

BE CAREFUL! Handle animals with care.

1. As a group, choose a Mealworm Observation Station that your teacher has set up. Each station has three jars labeled A, B, and C.

2. **Observe** Break into smaller groups. Each small group should observe the animals in one jar. Record your observations. Share your observations with the group.

_____

_____

3. Record all the questions you have about mealworms and how they change. How could you find the answers?

_____

4. **Experiment** Design simple experiments to find out more about the mealworms. Do they prefer light or dark places? What source of energy do they use to live and grow?

_____

5. **Observe** Make observations of the animals every few days. Record your observations.

_____

_____

©Macmillan/McGraw-Hill

Unit B · Animals as Living Things   Use with textbook page B55   59

Name_____ Date_____

Drawing Conclusions

1. **Communicate** Describe the stages of mealworm development.
   _____
   _____

2. Use your drawings to arrange the stages in the order in which you think mealworm develop.
   _____

3. **FURTHER INQUIRY** Infer How does the way a mealworm grows differ from how other animals like cats and dogs grow?
   _____
   _____

Inquiry

Think of your own questions that you might like to test. Do other insects go through changes as they grow?

My Question Is:
_____
_____

How I Can Test It:
_____
_____

My Results Are:
_____
_____
_____

60   Unit B · Animals as Living Things   Use with textbook page B55

Name	Date

Alternative Explore

Lesson 5

Changing Fruit Flies

Procedure

Materials
- jars containing food and fruit flies in different stages of development

1. As a group, observe a jar of fruit flies. Record your observations.

_____

_____

2. Make observations of the animals every few days. Record your observations. Draw the stages that you see.

_____

Drawing Conclusions

1. Describe the stages of development that you saw.

_____

_____

2. In what order do the stages occur?

_____

3. How does the way a fruit fly grows and develops differ from the way other animals, such as cats and dogs grow and develop?

_____

_____

© Macmillan/McGraw-Hill

Unit B · Animals as Living Things | **Use with TE textbook page B55** | 61

Name_____ **Date**_____

QUICK LAB
FOR SCHOOL OR HOME
Lesson 5

Heredity Cards

Hypothesize How many possible offspring can come from six different traits?

Write a **Hypothesis:**

_____

_____

Materials

- yellow construction paper
- green construction paper
- scissors
- marker

Procedure

1. Cut three cards from each paper. Yellow cards represent the female, green cards the male. Store your cards.

2. Write a trait for "Hair," "Eye color," and "Height" on one set of cards. Make sure the traits on the other set are different.

3. Match cards to make "offspring." Each offspring needs one card for each trait.

4. Continue matching cards to create offspring. Give each a number. Record the traits in a table. Use another piece of paper.

Drawing Conclusions

5. **Observe** How many different offspring did you get? _____

6. **Predict** How many offspring would you get with eight cards?

_____

_____

Unit B · Animals as Living Things

Use with textbook page B62

©Macmillan/McGraw-Hill

Name_____ **Date**_____

7. Going Further Think about a family you know in which there are several children. Is there one trait that all the children have? Is there a trait that only one of the children has? Write and conduct an experiment.

My Hypothesis Is:

_____

_____

My Experiment Is:

_____

_____

My Results Are:

_____

_____

©Macmillan/McGraw-Hill

Unit B · Animals as Living Things   Use with textbook page B62   63

Name_____ Date_____

Explore Activity
Lesson 6

How Can Body Color Help an Animal Survive?

Hypothesize What role does body color play in the types of places an animal can stay without being noticed?

Write a **Hypothesis:**
_____
_____

Materials
- colored toothpicks
- plastic bag or shoe box
- label or piece of masking tape
- marking pen

Procedure

1. Label your bag or box with your name. This is your "nest." Use it to hold all the toothpick "worms" that you collect.

2. **Observe** Follow the rules given by your teacher to capture the worms. Record the rules. Also record any observations that you make while collecting the worms.

   _____
   _____
   _____
   _____
   _____

3. **Communicate** When you are done, record your results in a bar graph like the one shown.

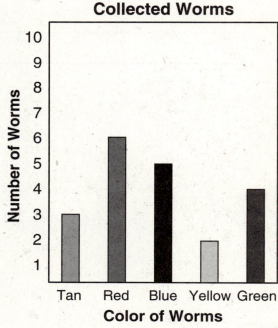

Collected Worms

64 Unit B · Animals as Living Things   Use with textbook page B65

Name_____ Date_____

Drawing Conclusions

1. Which color worms were easiest to see? Why?

   _____
   _____

2. Which color worms were hardest to see? Why?

   _____
   _____

3. If you were to become a toothpick worm, what color would you want to be? Why?

   _____
   _____

4. **FURTHER INQUIRY** Predict Colors help certain animals blend in with their surroundings. Why do you think some animals have bright colors? How could you find out? Try it and report your results.

   _____
   _____

Inquiry

Think of your own questions that you might like to test. Why do other animals change color?

My Question Is:

_____

How I Can Test It:

_____
_____

My Results Are:

_____
_____

Unit B · Animals as Living Things   Use with textbook page B65   65

Name_____ **Date**_____

Alternative Explore
Lesson 6

Designing Tigers

Procedure

1. Your teacher will assign your group a habitat, such as the desert, rainforest, African plain, pine forest, or arctic region.

2. Obtain a tiger shape. Discuss with your group the coloring and pattern that would best blend in with the habitat that you were assigned. (You don't have to use stripes.)

3. What color and pattern did you choose?

_____

4. Color your tiger to follow your group's design.

5. Show your tiger to another group. Discuss how well your tiger blends in with the other group's habitat. Would their tiger blend in with your habitat?

_____

Materials

• drawing of a tiger shape

• art materials

Drawing Conclusions

1. How did you decide on the color and pattern of your tiger?

_____

_____

_____

2. Would your tiger survive in the other group's habitat? Explain.

_____

_____

3. Why is an animal's coloring important?

_____

_____

© Macmillan/McGraw-Hill

66 Unit B · Animals as Living Things   Use with TE textbook page B65

Name_____ Date_____

Inquiry Skill Builder
Lesson 6

Form a Hypothesis

How Do Adaptations Help an Animal Survive?

Every science experiment begins with a hypothesis. A hypothesis is a statement you can test. "Dogs like big bones best" is a hypothesis. You could test this hypothesis by giving dogs different-sized bones.

In this activity you will design two different kinds of animals—a super predator and a prey animal that is skilled at avoiding predators. Then form a hypothesis about how their adaptations would help each animal in different situations.

Materials

- modeling clay
- construction paper
- drawing materials

Procedure

1. What traits should your predator have? Record them. Describe how these traits would help the animal.

   _____

   _____

   _____

2. Do the same for your prey animal.

   _____

   _____

   _____

3. **Communicate** Make a table like the one shown for each animal. Fill in each category that applies. Add any extra categories that you need.

4. **Make a Model** Make models or colored drawings of your animals. Label all the features of your animals. Tell how they function.

Animal Name _____
Predator ☐   Prey ☐
Food _____
Enemies _____
Environment _____

Trait	How it Helps
Length	
Weight	
Shape	
Coloring	
Pattern	
Skin	
Arms	
Legs	
Tails	
Fins	
Eyesight	
Hearing	
Smell	
Strength	
Quickness	
Intelligence	

© Macmillan/McGraw-Hill

Unit B · Animals as Living Things   **Use with textbook page B69**   67

Name_____ Date_____

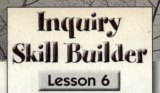

5. Form a Hypothesis How would these features help the animal survive?

_____

Drawing Conclusions

1. **Communicate** What are the animals' most important features? How would they use these features?

_____
_____
_____

2. Review your hypothesis. How could you test it?

_____
_____
_____

3. **Predict** Predict what would happen if you could test your hypothesis.

_____
_____
_____

68 Unit B · Animals as Living Things   Use with textbook page B69

Name_____ Date_____

Explore Activity
Lesson 1

How Can You Interpret Clues in Rocks?

Procedure

Materials

• 5 different rock samples

• hand lens

1. **Observe** Carefully observe each rock. Describe and record its properties on a separate piece of paper. Observe the rock's color, hardness, texture, and shininess. Is it made of similar particles that you can see? Does it have any layers?

2. **Observe** Use a hand lens to observe each rock sample. Record your observations on a separate piece of paper. Compare them with those of your classmates. Make a class list of all the properties you observed.

3. **Classify** Use the property words to make a rule that will help you divide the rocks into two groups. Make a new rule and try it again. Have others observe your two groups. See if they can guess the rule you used to group them.

_____

Drawing Conclusions

1. **Infer** Which rocks may have formed from sand or gravel? Which may have formed on an ocean bottom? What evidence supports your answers?

_____

_____

2. **Compare** samples. How are they alike? How are they different?

_____

_____

3. **FURTHER INQUIRY** **Infer** How do you think these rocks formed? Use your observations to explain why you think so.

_____

_____

Unit C · Earth and Beyond   Use with textbook page C5   69

Name_____ Date_____

Alternative Explore
Lesson 1

Observing Leaves

Procedure

1. Obtain a group of leaves. Look for ways they are alike and different.

2. Sort your leaves into two groups. What characteristic of the leaves did you use to sort them?

_____

_____

3. Exchange sorted leaves with those of another student. Were your classification systems alike or different? Explain.

_____

_____

Materials

• a variety of different types of leaves

Drawing Conclusions

1. How is classifying leaves like classifying rocks?

_____

_____

_____

2. When you sorted leaves, you may not have classified them in the same way other students did. Why is it important for scientists to agree on a system of classification, whether it's for leaves or for rocks?

_____

_____

_____

©Macmillan/McGraw-Hill

70   Unit C · Earth and Beyond   Use with TE textbook page C5

Name_____ Date_____

Identifying Minerals

Hypothesize How can you tell minerals apart?
Write a **Hypothesis:**

_____

_____

Procedure BE CAREFUL! Wear goggles.

1. Use tools and the table on the next page to identify five mineral samples.

2. Draw and name each mineral on a separate piece of paper. Write the properties of that mineral.

Materials

- 5 mineral samples, labeled 1 to 5
- hardness testers— copper penny, your fingernail
- streak plate
- goggles (for "Going Further")

Drawing Conclusions

3. Which properties helped you most to identify each mineral?

_____

_____

_____

4. **Going Further** Scientists also use an acid test to identify minerals. Limestone contains calcium carbonate, which reacts with acid to release carbon dioxide gas. Sedimentary rocks formed in ocean floors often contain limestone from seashells. Perform an acid test using vinegar, which contains acetic acid, and a seashell.

You may also test an egg. What do you think the eggshell is made of?

My Hypothesis Is:

_____

_____

My Experiment Is:

_____

_____

Unit C · Earth and Beyond   Use with textbook page C8   71

Name_____ **Date**_____

QUICK LAB
FOR SCHOOL OR HOME
Lesson 1

My Results Are:

_____

_____

_____

Mineral	Color	Luster	Streak	Hardness	Other
Pyrite	brassy yellow	shiny like a metal	greenish-black	not scratched by testers	looks like gold; breaks unevenly
Quartz	colorless, white, pink, purple	glassy	white	not scratched by testers	breaks unevenly
Mica	colorless, silvery, brown	may look glassy	white	scratched by fingernail	splits into thin sheets
Feldspar	yellow, white, gray, red, brown	glassy, pearly	white	not scratched by testers	splits easily in two directions
Calcite	colorless, white	glassy	white	scratched by copper and iron	splits in three directions

©Macmillan/McGraw-Hill

72   Unit C · Earth and Beyond   Use with textbook page C8

Name_____ **Date**_____

QUICK LAB
FOR SCHOOL OR HOME
Lesson 1

Observing Sediments

Hypothesize What will happen if you put some rock particles and sand in water? Which will settle first? Last? Do you think the size of the particles matters? Write a **Hypothesis:**

_____

_____

Procedure **BE CAREFUL!** Wear goggles.

1. **Make a Model** Put some gravel and sand in a jar of water. Cover it with a lid. Shake the jar. Set it aside.

Drawing Conclusions

2. Sketch what you see on a separate piece of paper. How many layers formed? Which layer settled first? Last?

_____

_____

3. **Infer** How does this illustrate the formation of sedimentary rocks?

_____

4. **Going Further** What happens when soil of different compositions mix? Write and conduct an experiment.

My Hypothesis Is:

_____

My Experiment Is:

_____

My Results Are:

_____

Materials

- gravel
- spoon
- fine-grained sand
- clear quart jar with a lid, half-filled with water
- goggles

Unit C · Earth and Beyond   Use with textbook page C11   73

©Macmillan/McGraw - Hill

Name_____ Date_____

Explore Activity
Lesson 2

What Can You Learn from Fossils?

Hypothesize How can you interpret clues left from millions of years ago? What types of things can you look for in fossils to learn the story they tell? Write a **Hypothesis:**

Materials
- footprint puzzle

_____

_____

Procedure

1. **Observe** Carefully study the footprints on page 76. Look for clues in the sizes and types of prints. Think about which were made first, next, and last.

2. **Communicate** Discuss the evidence with your partner. How can you work together to interpret it?

_____

3. Record the "story" you think the prints tell.

_____

_____

_____

Drawing Conclusions

1. **Infer** How many animals made the tracks? Are all the animals the same kind? How can you tell?

_____

_____

74   Unit C · Earth and Beyond   Use with textbook page C17

Name _____ **Date** _____

Explore Activity
Lesson 2

2. **Infer** Were all the animals moving in the same direction? How do you know? Which came first? Next? Last?

_____

_____

_____

_____

_____

3. How does your story compare with those of your classmates? On what points do you agree? Disagree? Be prepared to defend your interpretation.

_____

_____

4. **FURTHER INQUIRY** **Interpret Data** Create another footprint puzzle. Challenge a classmate to figure out the story the footprints tell.

Inquiry

Think of your own questions related to interpreting footprints. Can you identify the footprints of common animals?

My Question Is:

_____

How I Can Find Out:

_____

My Results Are:

_____

_____

©Macmillan/McGraw-Hill

Unit C · Earth and Beyond   Use with textbook page C17   75

Name_____ Date_____

Explore Activity
Lesson 2

76 Unit C · Earth and Beyond   Use with textbook page C17

Name_____ Date_____

Clay Stories

Procedure

Materials
- slab of clay

1. Obtain a slab of clay, and roll it to soften it.

2. Work in a group to make impressions in the clay. The impressions should tell a story. For example, you can make footprints by drawing them with a paper clip or other objects.

3. Exchange your group's finished clay story with that of another group. Describe the other group's clay story. What do you think the story is?
   _____
   _____
   _____
   _____

4. Tell the other group what you think their story is. Were you correct?
   _____

Drawing Conclusions

1. What kind of details did you use to interpret the other group's story?
   _____
   _____

2. How is what you did similar to the way scientists interpret fossils?
   _____
   _____

Unit C · Earth and Beyond   Use with TE textbook page C17

Name_____ Date_____

QUICK LAB
FOR SCHOOL OR HOME
Lesson 2

Making Molds and Casts

Procedure

1. **Make a Model** Coat a shell with petroleum jelly. Then firmly but gently press the shell into the clay.

2. Carefully remove the shell from the clay. Fill the clay with plaster of Paris.

3. When the plaster has dried, remove it from the clay.

Materials

- seashells
- petroleum jelly
- modeling clay
- container of plaster of Paris

Drawing Conclusions

4. Which is the mold? Which is the cast? How are they similar and different?

_____

_____

_____

_____

5. What shell characteristics can you see in the mold? In the cast?

_____

_____

6. **Going Further** What features do you think scientists observe in animal casts and molds? Write and conduct an experiment.

My Hypothesis Is: _____

_____

My Experiment Is: _____

_____

My Results Are: _____

_____

©Macmillan/McGraw-Hill

78   Unit C · Earth and Beyond   Use with textbook page C19

Name_____ Date_____

Inquiry Skill Builder
Lesson 2

Use Numbers

Dinosaur Bones

Footprint size gives a good idea of overall size and height. Scientists have determined that the length of a footprint is generally equal to one-quarter the length of the hind leg bone of the animal that made it. The length of the bone gives a good idea of the animal's overall size.

Procedure

1. **Collect Data** This table gives the footprint length of six adult dinosaurs.

A Name of Dinosaur	B Length of Footprint	C Probable Length of Hind Leg Bone	D Probable Rank in Overall Size
Triceratops	15 inches ($1\frac{1}{4}$ feet)		
Tyrannosaurus	30 inches ($2\frac{1}{2}$ feet)		
Stegosaurus	18 inches ($1\frac{1}{2}$ feet)		
Velociraptor	6 inches ($\frac{1}{2}$ foot)		
Compsognathus	3 inches ($\frac{1}{4}$ foot)		
Ultrasaurus	78 inches ($6\frac{1}{2}$ feet)		

2. **Use Numbers** Determine how to calculate the lengths of the hind leg bones. Complete Column C.

3. Rank the dinosaurs in order of probable overall size. Write 1 for the largest and 6 for the smallest in Column D.

Drawing Conclusions

1. **Interpret Data** Which dinosaur probably had the largest hind leg bone? The smallest? _____

2. Which two dinosaurs were probably close in size? The most different in size? _____

_____

©Macmillan/McGraw-Hill

Unit C · Earth and Beyond   Use with textbook page C23   79

Name _____ **Date** _____

Explore Activity
Lesson 3

How Do Glaciers Scratch and Move Rocks?

Hypothesize The main component of a glacier is ice. How can a block of ice help shape Earth's surface? Write a Hypothesis:

_____

_____

_____

Materials

- paper towel
- clean ice cube
- ice cube made with sand or gravel
- aluminum foil
- wood scrap

Procedure: Design Your Own

1. **Form a Hypothesis** Look at the two ice cubes. Which do you think will cause more changes as it moves across a surface? Record your answer, which is a hypothesis. Explain why you think it is correct.

   _____

   _____

2. **Experiment** Design an experiment to test your hypothesis. Use only the materials that your teacher provides. Record the results.

   _____

   _____

   _____

   _____

3. **Observe** Place the ice cubes on a folded paper towel. Allow them to melt. Observe and record what they leave behind.

   _____

   _____

80   Unit C · Earth and Beyond   Use with textbook page C33

Name _____ Date _____

Drawing Conclusions

1. How did each model feel as you rubbed it over a surface?
   _____
   _____

2. **Interpret Data** Did the observations support your hypothesis? Explain.
   _____
   _____

3. Describe what was inside each ice cube.
   _____
   _____

4. FURTHER INQUIRY **Infer** How would a heavier glacier affect how the land changed? Conduct a new experiment to answer the question and test your inference.
   _____
   _____

Inquiry

Think of your own questions related to glaciers. Are materials other than rocks trapped in glaciers? What happens to them?

My Question Is:
_____

How I Can Test It:
_____

My Results Are:
_____
_____

Unit C · Earth and Beyond   Use with textbook page C33

Name_____ **Date**_____

Alternative Explore
Lesson 3

Other Action

Procedure **BE CAREFUL!** Wear goggles.

Materials

- rocks and sand
- ice cube tray
- goggles
- wooden board

1. You will make a glacier by freezing rocks and sand with water in an ice cube tray. Decide whether you want to use mostly sand or rocks. Which do you think will make more scratches on a wooden board?

_____

_____

2. Put the rocks and sand in the ice cube tray. Make sure each space in the tray has the same amount of rocks and sand. Pour water into the tray. Let your glacier freeze overnight.

3. Try scratching a wooden board with your glacier. Did the glacier scratch the board as much as you expected? In the space below, draw the scratches.

4. Compare your results with those of other groups.

Drawing Conclusions

1. Did you use mostly rocks or mostly sand to make your glacier?

_____

2. After comparing results with other groups, compare the scratches made by rocks with the scratches made by sand.

_____

_____

82   Unit C · Earth and Beyond   Use with TE textbook page C33

©Macmillan/McGraw-Hill

Name _____ **Date** _____

Inquiry Skill Builder
Lesson 3

Define Terms

Flow of a Glacier

What do we mean when we say that a glacier flows? In this activity you will make and observe a model to see how glacial ice flows. Then you will be able to define glacial ice flow based on your experiences and observations.

Materials

- goggles
- prepared cornstarch mixture
- mixture of sand, gravel, and soil
- waxed paper
- metal spoon
- ruler

Procedure **BE CAREFUL!** Wear goggles.

1. **Make a Model** Place a spoonful of the cornstarch mixture on a piece of waxed paper. This represents a glacier. Record what happens.

   _____

   _____

2. **Observe** Place another spoonful on top of the first. This represents new snow. Record what happens.

   _____

   _____

3. Sprinkle some of the sand mixture in a 3-cm band around the edges and on top. Mark the edges of the sand on the waxed paper.

4. **Observe** One at a time, add four more spoonfuls of the cornstarch mixture. Each time, mark how far the glacier moves and the sand's position.

5. **Observe** Flip the glacier over onto another piece of waxed paper. Measure and draw the bottom.

© Macmillan/McGraw-Hill

Unit C · Earth and Beyond   Use with textbook page C36   83

Name_____ **Date**_____

Inquiry Skill Builder
Lesson 3

Drawing Conclusions

1. **Explain** Did the sand mixture sprinkled on top in step 3 eventually reach the bottom?

   _____

   _____

2. **Interpret Data** What do you think happens when a real glacier moves over rocks and boulders?

   _____

   _____

   _____

3. **Define Terms** Define *glacial ice flow*.

   _____

   _____

   _____

84   Unit C · Earth and Beyond   Use with textbook page C36

Name_____ **Date**_____

Explore Activity
Lesson 4

What Is Soil Made Of?

Procedure

1. Spread the newspaper on a desk or table. Place one soil sample on each paper towel. Put the paper towels on the newspaper.

2. **Observe** Use a pencil to push around the soil a little bit. Observe each sample with the hand lens. Record your observations of each soil sample.

_____

_____

3. **Classify** Use the pencil tip to classify the particles of each sample into two piles—pieces of rock and pieces of plant or animal material.

4. **Observe** Put four drops of water on each sample. After a few minutes, check which sample leaves the biggest wet spot on the newspaper.

Materials

- 3 types of soil
- hand lens
- eye dropper
- water
- newspaper
- paper towels
- 3 sharp pencils

Drawing Conclusions

1. **Infer** What kinds of materials make up each soil sample?

_____

2. How do the particles you sorted in each soil sample compare by size? By color? _____

3. **Observe** Describe the properties you observed of each sample.

_____

_____

4. Which sample absorbed the most water? How can you tell?

_____

5. **FURTHER INQUIRY** **Infer** How do you think soil is made? Use evidence to support your explanation.

_____

Unit C · Earth and Beyond | Use with textbook page C43 | 85

Name_____ Date_____

Alternative Explore
Lesson 4

Other Soils

Procedure

1. Collect a soil sample. Where did you collect it?

_____

_____

2. Place the soil sample on a paper towel, and place the paper towel on the newspaper.

3. Use a pencil to push the sample around a bit. Use the hand lens to get a closer look at the soil. Record your observations.

_____

_____

4. Use the pencil tip to separate the soil sample into pieces of rock and pieces of plant and animal material.

5. If you don't remember what the soil samples from the Explore Activity were like, examine them again.

Materials

- 3 types of soil
- eye dropper
- newspaper
- 3 sharp pencils
- hand lens
- water
- paper towels
- your own soil sample

Drawing Conclusions

1. How did your sample compare to the soil samples from the Explore Activity?

_____

_____

2. How do you think the content of your soil sample was affected by the place where you collected it?

_____

_____

_____

©Macmillan/McGraw-Hill

86   Unit C · Earth and Beyond   Use with TE textbook page C43

Name _____ **Date** _____

Rate of Flow

QUICK LAB
FOR SCHOOL OR HOME
Lesson 4

Procedure

1. Make two containers like the one shown on page C48 in your textbook. Put sandy soil in one container. Hold the container over a measuring cup. Slowly pour 1 cup of water over the soil, and start timing.

2. **Measure** When water drops begin to "hang," record the total time. Determine the amount of water left in the soil. Record your findings.

   _____

   _____

3. Repeat with the clay-rich soil in the other container.

Materials

- 2 prepared containers
- sandy soil
- clay-rich soil
- water
- 2 measuring cups
- stopwatch or clock with second hand

Drawing Conclusions

4. Through which soil did the water pass more quickly? Which soil allowed more water to pass? _____

   _____

5. **Interpret Data** Relate your findings to soil texture.

   _____

6. **Going Further** How would the addition of humus affect the water flow and absorption rates of the two sample soils? Conduct an experiment.

   My Hypothesis Is:

   _____

   _____

   My Experiment Is:

   _____

   My Results Are:

   _____

© Macmillan/McGraw-Hill

Unit C · Earth and Beyond   **Use with textbook page C48**   87

Name _____ Date _____

Explore Activity
Lesson 5

What's Inside?

Hypothesize How can you learn about something that you can't see directly? Write a **Hypothesis:**

_____

_____

Materials

- 3 sealed opaque containers with objects inside

Procedure: Design Your Own

1. What kinds of observations can you make about the objects in the containers? Make a plan with your group. Outline different things you can test. Record your plan. Compare your plan with another group's. Adjust your plan to make it better.

_____

_____

2. **Observe** Make your observations. Be sure you do not damage the containers. Each group member should have a turn with each container. Record all the observations.

_____

_____

3. **Interpret Data** Study your data. What clues do your observations provide?

_____

_____

88   Unit C · Earth and Beyond   Use with textbook page C53

©Macmillan/McGraw-Hill

Name_____ Date_____

Drawing Conclusions

1. **Infer** What do you think is in each container? On a separate page, draw a diagram or model that supports your observations.
   _____
   _____

2. **Communicate** Present your observations for each test. Explain how they support your conclusions.
   _____
   _____

3. FURTHER INQUIRY **Predict** Prepare a new mystery container for a partner. Do you think your partner will be able to guess what's inside? Provide a detailed written plan to be followed that will repeat your investigation. Present the results of your prediction.
   _____
   _____

Inquiry

Think of your own questions you might like to ask. What decisions might you make today based on something that you can't see or feel directly?

My Question Is:
_____

How Can I Test It:
_____
_____

My Results Are:
_____
_____

Unit C · Earth and Beyond   Use with textbook page C53   **89**

Name_____ Date_____

Hearing Evidence

Procedure

Materials
- objects that will make noise

1. Choose some objects that will make noise, such as beans in a container, chalk on the blackboard, or a book dropping on the floor. Don't tell your partner what objects you chose.

2. Have your partner sit with his or her eyes closed.

3. Make a noise. Ask your partner to identify what made the noise. What did your partner say the noise was?
   _____
   _____

4. Switch roles with your partner and repeat the experiment. Were you able to identify what made the noise your partner made?
   _____

Drawing Conclusions

1. What did your partner do to make a noise?
   _____
   _____

2. How did you identify the noise?
   _____
   _____
   _____

Unit C · Earth and Beyond   Use with TE textbook page C53

Name_____ Date_____

Earthquake Vibrations

Quick Lab
FOR SCHOOL OR HOME
Lesson 5

Hypothesize How do you think the energy of earthquakes travels through Earth? Write a **Hypothesis:**

_____

_____

_____

Materials

- marble
- pan of water
- newspaper
- flashlight

Procedure

1. Spread out some newspaper to absorb splashed water. Place a pan of water on the newspaper.

2. **Observe** Take turns dropping a marble into the water from a height of about 15 cm (6 in.). Shine a flashlight on the water to see more clearly. Record your observations of each wave pattern.

_____

_____

_____

Drawing Conclusions

3. **Communicate** What wave pattern did the marble create?

_____

_____

4. **Infer** How do you think this pattern might relate to the way earthquake vibrations travel?

_____

_____

Unit C · Earth and Beyond   Use with textbook page C55   91

Name_____ Date_____

QUICK LAB
FOR SCHOOL OR HOME
Lesson 5

5. **Going Further** When you raise the marble over the water, you are working against gravity and using energy. This energy is stored in the marble. When the marble is released and drops into the water, this energy is used to move the water. Raising the marble higher, or lifting a heavier marble, requires more energy, so the marble releases more energy as it falls into the water. How would you expect the wave size to change as a marble is dropped from different heights? How would the wave size change if marbles with different weights were dropped from the same height? Relate this information to earthquakes. Write and conduct an experiment.

My Hypothesis Is:

_____

_____

My Experiment Is:

_____

_____

_____

_____

My Results Are:

_____

_____

_____

_____

_____

© Macmillan/McGraw-Hill

92   Unit C · Earth and Beyond   Use with textbook page C55

Name_____ Date_____

Explore Activity
Lesson 6

How Do the Sun, Earth, and the Moon Move?

Hypothesize How will moving Earth and Moon models around a lamp change the appearance of the Moon model? Write a **Hypothesis:**

_____
_____

Materials
- lamp without a shade
- 75 or 100-watt bulb
- extension cord
- craft-foam ball

Procedure

1. **Observe** Look at the pictures that show the ways the Moon appears from Earth at night.

2. **Make a Model** Create a model of the Sun, Earth, and the Moon. Let the lamp be the Sun. One partner should hold the ball to model the Moon. The other partner should represent Earth.

3. **Experiment** Turn out the lights in the room. Turn on the lamp. Do not move the lamp. Experiment with different positions to try to model the different ways the Moon looks in the pictures.

4. **Make a Model** Again let the lamp be the Sun. This time let the ball be Earth. Do not move the lamp. Demonstrate and explain what you think causes day and night on Earth.

Drawing Conclusions

1. What causes the Moon to look different to us on Earth from night to night?

_____
_____
_____

Unit C · Earth and Beyond   Use with textbook page C65   93

Name_____ Date_____

2. Why does Earth have day and night?
_____
_____

3. **Infer** Why are the patterns of day and night and changes in the Moon's appearance so predictable? _____
_____
_____

4. You can model day and night on Earth by keeping the ball still and moving the lamp. This is not an accurate model. Why?
_____
_____

5. FURTHER INQUIRY **Infer** How can you change the model so that the moon phases are opposite in order? Try it to test your inference.
_____
_____

Inquiry

Think of your own questions that you might like to test. What causes different seasons?

My Question Is:
_____
_____

How I Can Test It:
_____
_____

My Results Are:
_____
_____
_____

Unit C · Earth and Beyond   Use with textbook page C65

Name_____ Date_____

Alternative Explore
Lesson 6

Modeling the Sun, Earth, and the Moon

Materials
- index cards with the names Earth, Moon, and Sun written on them

Procedure

1. One student should be the "Sun" and stand in place. This student should hold the index card with the name Sun high above her or his head.

2. A second student should model Earth's movement in space. This student should hold the index card with the name Earth in front of her or him. Then the student should slowly turn completely around. At the same time, the student should also slowly circle the "Sun."

3. A third student should model the Moon's movements in space. This student should hold the index card with Moon in front of her or him. At the same time, the student should slowly circle "Earth," always facing it.

Drawing Conclusions

1. How does Earth move?
   _____
   _____
   _____

2. What part of Earth experiences day? What part experiences night?
   _____
   _____
   _____

3. How does the Moon move?
   _____
   _____
   _____

Unit C · Earth and Beyond   Use with TE textbook page C65   95

Name_____ Date_____

Inquiry Skill Builder
Lesson 6

Interpret Data

Moon Phases on a Calendar

When you interpret data, you use information from a picture, a table, or a graph. A calendar is a type of table. Each icon in this calendar tells you what phase the Moon will be in for that day. Interpret the data in the calendar to answer the questions on the following page.

SEPTEMBER						
Sunday	Monday	Tuesday	Wednesday	Thursday	Friday	Saturday
		1	2	3	4	5
6	7	8	9	10	11	12
13	14	15	16	17	18	19
20	21	22	23	24	25	26
27	28	29	30			

96   Unit C · Earth and Beyond   Use with textbook page C74

Name _____ **Date** _____

Inquiry Skill Builder
Lesson 6

Procedure

1. **Communicate** Make a table that shows how many times each phase of the Moon appears in the month shown on this calendar.

2. Make another table listing each phase of the Moon discussed on pages C72 and C73 in your textbook. Next to each phase, write the day of the month that phase occurs for the month shown on this calendar.

3. **Interpret Data** On which day or days was there a new Moon? A first quarter Moon? A gibbous Moon?

   _____

   _____

4. **Interpret Data** Were there any days this month that had the same phase of the Moon? If so, what were they?

   _____

Drawing Conclusions

1. **Observe** What pattern do you see in the phases of the Moon for this month?

   _____

   _____

2. **Interpret Data** Find a calendar that shows the phases of the Moon. Compare the month shown in this activity with one month in your calendar. How are the phases of the Moon similar? Different?

   _____

   _____

Unit C · Earth and Beyond   Use with textbook page C74   97

	Explore Activity
Name_____ **Date**_____	Lesson 7

How Do Objects in the Night Sky Compare in Size?

Hypothesize How do the size of the Sun, the Moon, and the planets compare to each other? Write a **Hypothesis:**

_____

_____

_____

_____

Materials

- colored craft paper
- sheets of newspaper
- meterstick
- metric ruler
- marker
- scissors

Procedure **BE CAREFUL!** Handle scissors carefully.

1. **Measure** Study the table. Compare diameters of different objects in the night sky. The diameter is the distance across the middle of a circle or sphere.

Comparing Diameters	
Object	**Size** (in Earth Diameters)
Earth	1
Moon	$\frac{1}{4}$
Mars	$\frac{1}{2}$
Saturn	$9\frac{1}{2}$
Jupiter	11
Sun	109

2. **Make a Model** Make a paper circle 1 cm across to model Earth. Measure and cut paper circles to model each object listed in the table. If you cannot make the model Sun large enough, make it as large as possible. Label each object.

3. **Classify** Arrange the objects in a way that lets you compare their sizes.

© Macmillan/McGraw-Hill

98   Unit C · Earth and Beyond   Use with textbook page C79

Name_____ Date_____

Drawing Conclusions

1. Compare the sizes of the Moon, the Sun, and the planets.
   _____

2. How can the Moon and the Sun look the same size in our sky?
   _____
   _____

3. **Infer** Our Sun is an average-sized star. Why do other stars look so much smaller than the Sun?
   _____
   _____

4. FURTHER INQUIRY **Make a Model** Use paper to make a model of the Sun and its planets. Use your model to answer questions you have about the solar system.

Inquiry

Think of your own questions that you might like to test. How far away are objects that can be seen in the night sky?

My Question Is:
_____

How I Can Test It:
_____
_____

My Results Are:
_____
_____

Unit C · Earth and Beyond   Use with textbook page C79   99

Name _____ **Date** _____

Alternative Explore
Lesson 7

The Night Sky

Procedure

1. Tape the 10-cm circle to a wall. Now stand 3 meters away from the wall and hold up the 5-cm circle.

2. Which circle appears larger? Why?

   _____

   _____

   _____

3. How does distance affect an object's appearance?

   _____

   _____

Drawing Conclusions

1. Mars is about twice as large as the Moon. Why does Mars appear much smaller in the sky than the Moon?

   _____

   _____

2. Saturn is about 18 times larger than Mars. Why does the Moon appear larger in the sky than Saturn?

   _____

   _____

Materials

- 2 paper circles: 5 cm and 10 cm

©Macmillan/McGraw-Hill

100   Unit C · Earth and Beyond   Use with TE textbook page C79

Name_____ Date_____

A Comet's Tail

Hypothesize How does the tail extend from a comet? Write a **Hypothesis**:
_____
_____

Materials
- clay or small ball
- strips of tissue paper
- pins
- book or folder

Procedure

1. Use a small ball, or make one out of clay, to model a comet. Attach a few strips of tissue paper to the model. This will be the tail of the comet.

2. Go outdoors if it is windy, or have your partner wave a notebook or folder to create wind.

3. Hold the comet in the wind in different positions. Try to make the tail move in the wind.

Drawing Conclusions

4. Compare your model to a real comet. How might a comet's tail move?
_____
_____

Unit C · Earth and Beyond   Use with textbook page C87   101

Name_____ Date_____

5. **Going Further** A comet's tail forms as the Sun melts the materials in it. How could you use ice and a hair dryer to show how a comet's tail grows? What changes do you think you will see as you do the experiment?

My Hypothesis Is:

_____

_____

My Experiment Is:

_____

_____

My Results Are:

_____

_____

Name _____ **Date** _____

Explore Activity
Lesson 1

Where Can Water Be Found?

Hypothesize Imagine a stegosaur drinking from a pond millions of years ago. What happened to the pond? Write a **Hypothesis:**

_____

_____

Materials

- 6 different-colored markers
- white drawing paper

Make a water path to find out about the state of water in different places.

Procedure

1. Form six teams. Place one team at each location. Record your location. What is the state of the water there? Remember, the states of matter are solid, liquid, and gas.

   _____

2. Have each team go to the next closest location. Record the location and the state of the water. Repeat until every team has visited all six locations.

3. How did the state of water differ from location to location?

   _____

   _____

4. **Interpret Data** Use your color markers to draw your team's water path.

©Macmillan/McGraw-Hill

Unit D · Water and Weather   **Use with textbook page D5**   103

Name_____ Date_____

Drawing Conclusions

1. **Predict** Where might water stay in one place for a short time? A long time? Why?

   _____
   _____
   _____

2. Do you think water that was around at the time of the dinosaurs can still be around today? Why or why not?

   _____
   _____
   _____

3. FURTHER INQUIRY **Infer** What might have caused the change in the state of water from place to place? What evidence can you use to support your idea?

   _____

Inquiry

Think of your own questions that you might like to test. How does water return to the atmosphere?

My Question Is:

_____

How I Can Test It:

_____
_____

My Results Are:

_____

104 | Unit D · Water and Weather | Use with textbook page D5

Name_____ Date_____

Drinking with the Dinosaurs

Procedure

1. Look at the poster of the water cycle. With your group, locate the following places where water can be found:

    oceans and seas   plants and animals
    air and clouds   ice caps and glaciers
    soil and water   rivers and lakes

- poster of the water cycle without labels

2. Discuss with your group where places for water from the dinosaur age could still be around. List your group's ideas.

    _____
    _____
    _____

3. Look at your list of ideas. Discuss with your groups which of these locations hold water in place for a long time and which hold water for only a short time.

Drawing Conclusions

1. Where could you find water from the dinosaur age? Explain your answer.

    _____
    _____
    _____

2. Which locations hold water for a long time?

    _____
    _____

3. Which locations hold water for a short time?

    _____

Unit D · Water and Weather   Use with TE textbook page D5   105

Name_____ Date_____

QUICK LAB
FOR SCHOOL OR HOME
Lesson 1

Water in an Apple

Hypothesize Write the amount of water you think is in an apple. Write a **Hypothesis:**

_____

_____

Materials

- apple slices
- tray
- pan balance

Procedure

1. **Measure** Measure and record the mass of some apple slices.

   _____

2. Lay the apple slices on the tray, and put the tray in a warm place.

3. **Measure** When the slices are completely dried, measure their mass. Record the mass of the dry slices.

   _____

   _____

Drawing Conclusions

4. **Measure** What was the mass of the apple slices before and after drying?

   _____

5. Which had a greater mass—the fresh apple slices or dry apple slices? Why?

   _____

   _____

6. **Use Numbers** How much of the apple's mass was water? How did you find out?

   _____

   _____

106   **Unit D · Water and Weather**   Use with textbook page D12

Name_____ **Date**_____

QUICK LAB
FOR SCHOOL OR HOME
Lesson 1

7. **Going Further** What other fruits have water?
Write and conduct an experiment.

My Hypothesis Is:

_____

My Experiment Is:

_____

_____

My Results Are:

_____

_____

© Macmillan/McGraw-Hill

Unit D · Water and Weather   **Use with textbook page D12**   **107**

Name_____ Date_____

Explore Activity
Lesson 2

What Makes Water Disappear?

Hypothesize Where does puddle water go? How long does it take a small puddle to disappear? A big puddle? What causes the water to disappear? Write a **Hypothesis:**

_____

_____

Materials

- measuring cup
- water
- 2 index cards
- 2 lunch trays with sides

Procedure

1. **Measure** Pour a half cup of water into each tray.

2. Place one tray in a sunny area. Place the other in a dark area.

3. Use an index card for each tray. Label one card Sunny and the other card Dark. On each index card, write your name and the date. Then write the time when you placed the tray in its area.

4. **Observe** Check the trays every hour until the water in each is gone. Note on the appropriate index card how long it took the "puddle" to disappear. Record your results in the table below.

Time	Observation of Sunny Tray	Observation of Dark Tray

©Macmillan/McGraw-Hill

Unit D · Water and Weather   Use with textbook page D15

Name_____ Date_____

Drawing Conclusions

1. Which puddle disappeared first? Which took the longest to disappear?
   _____
   _____

2. What do you think made one puddle disappear faster? The other disappear more slowly?
   _____
   _____

3. FURTHER INQUIRY Experiment How might the variable of wind affect how fast water evaporates? Conduct an experiment to answer the question. Use evidence to support your explanation.
   _____
   _____

Inquiry

Think of your own questions that you might like to test using the water and lunch trays. What will happen to the water if the tray is covered?

My Question Is:
_____
_____

How I Can Test It:
_____
_____

My Results Are:
_____

Unit D · Water and Weather   Use with textbook page D15   109

Name_____ Date_____

Alternative Explore
Lesson 2

Salt Search

Procedure

1. Fill a cup halfway with water. Add a teaspoon of salt to the water. Stir the water until the salt dissolves.

2. Discuss with your group ways to get the salt back without heating the water. List your group's ideas.

_____

_____

Materials

- plastic drinking cup
- water
- salt
- plastic spoon

3. As a group, decide on a method you would like to try. Show your plan to your teacher before proceeding with it.

4. Try your plan. Record your results.

5. Share your plan and your results with the class.

Drawing Conclusions

1. How well did your plan work?

_____

_____

2. Which plan of your classmates worked best?

_____

_____

©Marmillan/McGraw-Hill

Unit D · Water and Weather   Use with TE textbook page D15

Name_____ Date_____

QUICK LAB

FOR SCHOOL OR HOME

Lesson 2

Disappearing Water

Hypothesize What happens when a glass of water is left uncovered? Write a **Hypothesis:**

_____

_____

Materials

- water
- 2 plastic cups
- piece of clear plastic wrap
- rubber band
- marker

Procedure

1. Fill two plastic cups halfway with water. Cover one cup with plastic wrap. Use a rubber band to hold it in place. Mark the water level water in each cup.

2. Place both cups in a warm, sunny spot.

3. **Predict** What do you think will happen in each cup?

_____

_____

4. **Observe** Check the cups every hour. Record what you see.

_____

_____

©Macmillan/McGraw-Hill

Unit D · Water and Weather   Use with textbook page D18   111

Name_____ Date_____

Drawing Conclusions

5. Where did the water in each cup go?
   _____
   _____

6. **Infer** Why did this happen?
   _____
   _____
   _____
   _____

7. **Going Further** What would happen if you repeated the experiment using ice cubes instead of water? Write and conduct an experiment.

   My Hypothesis Is:
   _____
   _____

   My Experiment Is:
   _____
   _____

   My Results Are:
   _____
   _____

Name_____ Date_____

Explore Activity
Lesson 3

What Makes the Ocean Move?

Hypothesize How could a message in a bottle travel across the ocean? Write a **Hypothesis:**

_____

_____

Procedure

BE CAREFUL! Handle the warm water carefully. Wear goggles.

1. Fill the box three-fourths full of room-temperature tap water.

2. Put the rocks in the bag. Fill the bag half full of warm water. Close it with the twist tie.

3. Place the bag in one corner of the box.

4. Float an ice cube in the opposite corner from the bag. If the ice cube melts, replace it.

5. Put food coloring in the dropper. Place four drops of the coloring in the water next to the ice cube.

6. **Observe** Look at the food coloring for several minutes through the sides of the box. Record what you see.

_____

_____

_____

Materials

- clear-plastic shoe box
- room-temperature tap water
- 500 mL of warm tap water
- small plastic sandwich bag
- twist tie
- food coloring
- dropper
- small rocks
- 2 or 3 ice cubes
- goggles

Unit D · Water and Weather Use with textbook page D27 113

Name_____ Date_____

Drawing Conclusions

1. **Observe** Where did the water sink? Where did it rise?
   _____
   _____

2. Why do you think you added food coloring to the water?
   _____
   _____

3. **FURTHER INQUIRY** **Experiment** Suppose you changed the activity. Predict what would happen if you did not use the bag of warm water. Try it. Do the results support your prediction? Explain.
   _____
   _____

Inquiry

Think of your own questions that you might like to test using the shoe box, water, and food coloring. What are other factors that could affect the way ocean water moves?

My Question Is:
_____
_____

How I Can Test It:
_____
_____

My Results Are:
_____

114 Unit D · Water and Weather   Use with textbook page D27

Name _____ **Date** _____

Alternative
Explore
Lesson 3

Rising Water

Procedure

BE CAREFUL! Wear goggles.

1. Half fill 2 cups with warm water. Add several drops of food coloring and stir.

2. Half fill 2 cups with cold water.

3. Pour one of the cups of warm water into one of the cups of cold water. Observe what happens. Record your observations.

_____

_____

_____

4. Pour some cold water into the remaining cup of warm water. Observe what happens. Record your observations.

_____

_____

_____

Materials

- 4 plastic drinking cups
- warm water
- food coloring
- plastic spoon
- cold water
- goggles

Drawing Conclusions

1. What happens when warm water and cold water are poured together?

_____

_____

2. What does this tell you about the difference between warm water and cold water?

_____

_____

© Macmillan/McGraw-Hill

Unit D · Water and Weather   Use with TE textbook page D27   115

Name_____ **Date**_____

QUICK LAB

FOR SCHOOL OR HOME

Lesson 3

Make Waves!

Hypothesize How does the wind affect waves?
Write a **Hypothesis:**

_____

_____

Materials

- clear-plastic shoe box
- water
- 2 straws
- cork

Procedure

1. Fill a plastic shoe box halfway with water. Place a cork at one end of the box.

2. **Predict** What will happen to the cork if you puff air at the water? What if you puff harder? At difference distances?

_____

_____

3. **Observe** Use the straw to puff at the water. Record what happens to the cork.

_____

4. **Experiment** Puff harder and at different distances. What happens to the cork each time?

_____

_____

©Macmillan/McGraw-Hill

116   Unit D · Water and Weather   Use with textbook page D32

Name _____ Date _____

Drawing Conclusions

5. **Communicate** Draw a diagram showing the cork's movements.

6. **Going Further** Describe how the cork moves if you and your partner both puff through a straw from opposite ends of the box. Write and conduct an experiment.

My Hypothesis Is:
_____
_____

My Experiment Is:
_____
_____
_____

My Results Are:
_____
_____

Name_____ Date_____

Explore Activity
Lesson 4

How Fast Does Water Flow in Soil and Rocks?

Hypothesize How do you think water travels through soil and rocks? Write a **Hypothesis:**

_____

_____

Procedure

1. With a pencil tip, make a small hole in the bottom of one paper cup.

2. Place your finger over the hole. Fill the cup with perlite or soil. Hold the cup over a plastic container. Have your partner pour in water to cover the perlite or soil.

3. **Observe** Take away your finger. Time how long it takes the water to drain. Record the results.

_____

_____

_____

4. Repeat using marbles to represent rocks.

_____

_____

Materials

- cup of perlite or soil

- cup of marbles

- two 12-oz paper cups

- pencil

- stopwatch

- plastic container

- 1 L of water

- measuring cup

118   Unit D · Water and Weather   Use with textbook page D37

©Macmillan/McGraw-Hill

Name_____ Date_____

Drawing Conclusions

1. Which material let water soak through faster?
   _____

2. How does the kind of material affect how fast water flows through it?
   _____
   _____

3. **Infer** What happens to rainwater falling on soil?
   _____
   _____

4. **FURTHER INQUIRY** **Infer** How fast will water flow through gravel compared with soil or rocks (marbles)? Make a prediction and then try it. Do the results support your prediction? Explain.
   _____
   _____
   _____

Inquiry

Think of your own questions that you might like to test using the water, paper cups, and other materials. How fast would the water flow through other soil types?

My Question Is:
_____

How I Can Test It:
_____
_____

My Results Are:
_____

Unit D · Water and Weather   Use with textbook page D37

Name_____ Date_____

Alternative Explore
Lesson 4

Soil and Growth

Procedure

1. Fill two pots with each type of soil.

2. Place a plant in each pot.

3. Place all plants in the Sun. Water all pots with the same amount of water every other day. Always use the same amount of water, no matter how wet or dry the soil or sand becomes.

4. Observe the growth of the plants for two weeks. Record your observations.

_____

_____

_____

Materials

- soil that is mostly clay
- potting soil
- sand
- 6 pots and saucers
- 6 identical plants

Drawing Conclusions

1. Which pots tended to be the driest?

_____

2. Which pots tended to be the wettest?

_____

3. In which soil did the plants grow best?

_____

_____

©Macmillan/McGraw-Hill

120   **Unit D · Water and Weather**   **Use with TE textbook page D37**

Name_____ Date_____

Make Runoffs

QUICK LAB — FOR SCHOOL OR HOME — Lesson 4

Hypothesize How do different soils affect runoff?
Write a **Hypothesis:**

_____
_____
_____
_____

Materials
- two 1-qt milk cartons
- plastic tray with sides
- soil
- sand
- marker
- 1 L of water
- measuring cup
- scissors

Procedure

BE CAREFUL! Handle scissors carefully.

1. Cut two milk cartons as shown. Label one Soil and the other Sand. Place them on the tray.

2. **Use Variables** Put an equal amount of sand and soil in the cartons. Draw the amounts on a separate piece of paper.

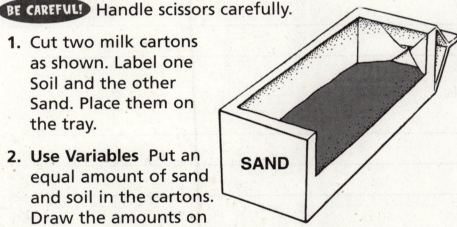

3. Fill a measuring cup with water. Slowly pour it over the soil until the soil can hold no more water. Determine and record the volume of water you poured into the soil.

_____
_____

4. Repeat step 3 for the sand.

Unit D · Water and Weather   Use with textbook page D39

| Name | Date |

QUICK LAB
FOR SCHOOL OR HOME
Lesson 4

Drawing Conclusions

5. Measure Which absorbed the most water? Which had the most runoff?

_____

_____

6. Going Further You inherit a map with an X marking where buried treasure is located. When you arrive at the spot, you discover it is near the ocean. Would you hope that the treasure was buried under dry or wet soil? Why? Write and conduct an experiment.

My Hypothesis Is:

_____

_____

My Experiment Is:

_____

_____

_____

My Results Are:

_____

_____

©Macmillan/McGraw-Hill

Unit D · Water and Weather   Use with textbook page D39

Name_____ Date_____

Inquiry Skill Builder
Lesson 4

Use Variables

Surface Area and Evaporation

Variables are things, or factors, in an experiment that can be changed to find answers to questions. In this activity, you'll answer this question: Does size or surface area of a puddle affect how fast it will evaporate? For a fair test, all of the factors in the experiment except the variable must remain the same. The only variable is surface area.

Procedure

BE CAREFUL! Handle scissors carefully.

1. **Make a Model** Use the sponges to make models of puddles with different surface areas.

2. **Measure** Place one sponge in each pan of the balance. Add paper clips to the pan with the smaller sponge until both sides of the balance are equal in mass.

3. **Infer** Find a way to add equal amounts of water to both sponges.

4. **Observe** Once you have set up your models, turn on the lamp. Check the models every half-hour. Record your observations in the table below.

Materials

- water
- measuring cup
- spotlight lamp
- small box of paper clips
- whole kitchen sponge
- half kitchen sponge
- scissors
- pan balance

Time	Observations

©Macmillan/McGraw-Hill

Unit D · Water and Weather   Use with textbook page D44   123

Name_____ Date_____

Inquiry Skill Builder
Lesson 4

Drawing Conclusions

1. **Infer** Which model became lighter first? What does this tell you about surface area and evaporation?

_____

_____

_____

2. **Identify** What variables did you change? Keep the same?

_____

_____

_____

3. **Experiment** What could you do to make water evaporate faster? Slower? Test your ideas.

_____

_____

_____

©Macmillan/McGraw-Hill

124   Unit D · Water and Weather   Use with textbook page D44

Name_____ Date_____

Explore Activity
Lesson 5

How Much Fresh Water Is Used?

Hypothesize People use—and waste—fresh water every day. How can you find out how much water is used at your school each day? Write a **Hypothesis:**

_____

_____

_____

Materials

- two 9-oz plastic cups
- measuring cup
- stopwatch, clock, or watch with second hand
- calculator (optional)

Procedure: Design Your Own

1. Determine ways to measure or estimate the amount of water used daily in school.

_____

_____

2. How can you figure out how much water is being used by each student? By each class? Record your results.

_____

_____

_____

3. **Communicate** Design a table to record all the data you gathered from your investigation.

© Macmillan/McGraw-Hill

Unit D · Water and Weather   **Use with textbook page D47**   125

Name_____ Date_____

Drawing Conclusions

1. Which activities used the most water each day? Which used the least?
   _____
   _____

2. **Use Numbers** How can you estimate how much water is used in the whole school in a day?
   _____
   _____

3. **Infer** From your observations can you think of ways to save water?
   _____

4. FURTHER INQUIRY **Communicate** Write a plan to measure or estimate the amount of water used daily at home. Record the data and present the results in a new table.
   _____
   _____

Inquiry

Think of your own questions that you might like to test. Where else is water used in large amounts?

My Question Is:
_____

How I Can Test It:
_____
_____

My Results Are:
_____

Unit D · Water and Weather   Use with textbook page D47

Name_____ Date_____

Alternative Explore

Lesson 5

Washing Hands

Procedure

1. Work within a group of 4 students. Place the plastic basin in the sink. Wash your hands, collecting all the water used in the basin.

2. Pour the water into the liter measure and record the amount.

3. Have each member of your group repeat step 2.

4. Find the average amount of water used to wash hands.

5. Ask each group member how many times he or she washes his or her hands each day.

6. Find the average number of times hands are washed each day.

Materials

- sink
- plastic basin
- liter measure

Student	Water Used	Number of Washes per Day
1		
2		
3		
4		
Average		

Drawing Conclusions

1. Round the average number of washes per day to the nearest whole number. Multiply the water used by the rounded number of washes per day to find the amount of water an average person uses for hand washing each day.

_____

2. What affects how much water is used each time you wash your hands?

_____

_____

Unit D · Water and Weather   Use with TE textbook page D47

© Macmillan/McGraw-Hill

Name_____ Date_____

QUICK LAB
FOR SCHOOL OR HOME
Lesson 5

Wasted Water

Hypothesize Can you estimate how much water a leaky faucet might waste in a day? Write a **Hypothesis:**

_____

_____

_____

Materials

- 1,000-mL (1-L) pitcher
- water faucet
- clock or watch with second hand
- calculator (optional)

Procedure

1. Turn on a faucet until it drips slowly.

2. **Predict** How much water do you think will be wasted in five minutes?

_____

3. Place a pitcher under the faucet for five minutes.

4. **Measure** Measure the collected water with a measuring cup.

_____

_____

Drawing Conclusions

5. **Observe** How much water was wasted in five minutes?

_____

6. **Use Numbers** If the faucet dripped like this every day, how much water would be wasted in an hour? In a day? In a week? In a year?

_____

_____

_____

©Macmillan/McGraw-Hill

Unit D · Water and Weather   Use with textbook page D54

Name_____ Date_____

7. **Going Further** If the faucet has dripped like this every day, how much water would have been wasted since you were born? Write and conduct an experiment.

My Hypothesis Is:

_____

_____

My Experiment Is:

_____

_____

My Results Are:

_____

Unit D · Water and Weather   Use with textbook page D54

Name_____ **Date**_____

Explore Activity
Lesson 6

What Can Air Do?

Hypothesize How does a parachute work?
Write a **Hypothesis:**

_____

_____

Procedure: Design Your Own

BE CAREFUL! Handle scissors carefully.

1. **Make a Model** Use a square from the plastic bag and string to make a parachute for a washer. Tie four pieces of string to the washer. Tie the other ends of the strings to the corners of the bag.

2. **Observe** Drop the parachute. Observe how the plastic bag changes as it falls.

3. **Experiment** Hold the parachute in one hand and the second washer in the other hand. Let them go at the same time. Observe both as they fall to the floor.

Materials

- large plastic trash bags cut in 12" × 12" pieces
- string
- scissors
- two washers

Drawing Conclusions

1. How did the plastic bag change when you dropped the parachute?

_____

_____

2. How does adding a parachute change the way a washer falls?

_____

_____

©Macmillan/McGraw-Hill

130   Unit D · Water and Weather   Use with textbook page D63

Name_____ **Date**_____

Explore Activity
Lesson 6

3. **Infer** Is air real? Use what you observed in this activity to support your answer.

_____

_____

4. **FURTHER INQUIRY** **Use Variables** What variables affect how fast the parachute falls? How can you change your parachute to make it fall more slowly? Try it and report your results.

_____

_____

Inquiry

Think of your own questions that you might like to test. Do paper bags or cloth parachutes give different results?

My Question Is:

_____

_____

How I Can Test It:

_____

_____

My Results Are:

_____

_____

©Macmillan/McGraw-Hill

Unit D · Water and Weather   Use with textbook page D63   131

Name_____ Date_____

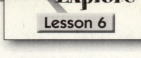

Alternative Explore
Lesson 6

Spinners

Procedure

BE CAREFUL! Handle scissors carefully.

Materials
- scissors
- paper

1. Cut a strip of paper about 8 inches by 2 inches. Hold it up and let it drop to the floor. What happened?

   _____

   _____

2. Starting at one end of the paper strip, cut going toward the middle of the strip. Cut about 4 inches into the strip.

3. Fold each of the new narrower strips in opposite directions, at a 90-degree angle to the uncut portion of the main strip.

4. Stand where there are no air currents. Hold the spinner with the two folded strips facing the floor, and drop it.

Drawing Conclusions

1. What happened when you dropped the spinner? Why?

   _____

   _____

2. Change the length of the "propellers" or change the weight of the paper. What happens now?

   _____

132 Unit D · Water and Weather   Use with TE textbook page D63

Name_____ **Date**_____

Inquiry Skill Builder
Lesson 6

Infer

Homemade Barometer

What happens to the air inside a container if the container's size changes? In this activity you will make a *barometer* (buh•RAHM•i•tuhr), a device for measuring air pressure. You will infer how a barometer measures changing air pressure. When you infer, you use observations to form an idea.

Procedure

BE CAREFUL! Handle scissors carefully.

1. **Make a Model** Use the scissors to cut the neck off one of the balloons. Stretch the bottom of the balloon over the mouth of the small jar. Hold it in place with a rubber band.

2. Glue the large end of the toothpick to the stretched balloon. Let the glue dry. Put the small jar inside the wide-mouthed jar.

3. Cut the neck from the second balloon. Stretch the bottom of the balloon over the mouth of the large jar. Hold it in place with a rubber band.

4. **Observe** Pull up and push down on the balloon stretched over the large jar. What happens to the toothpick on the small jar?

_____

Materials

- small jar, such as a baby food jar
- 1-qt or 1-L wide-mouthed jar
- two 9-in. balloons
- glue
- scissors
- flat toothpick
- 2 rubber bands

© Macmillan/McGraw-Hill

Unit D · Water and Weather   **Use with textbook page D70**   133

Name_____ Date_____

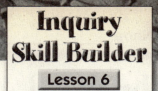

Drawing Conclusions

1. **Infer** How did the air pressure inside the large jar change when you pushed down on the balloon? What evidence helped you determine this?

   _____

   _____

2. **Infer** How did the air pressure inside the large jar change when you pulled up on the balloon? What evidence helped you determine this?

   _____

   _____

Name_____ Date_____

Explore Activity
Lesson 7

How Do Raindrops Form?

Hypothesize What will happen if you use ice to cool moisture in a jar? Write a **Hypothesis:**

_____

_____

Materials

- 1-qt or 1-L jar with lid
- ice cubes
- water

Procedure

1. **Make a Model** Pour water into the jar so that the bottom of the jar is covered.

2. Turn the jar lid upside down, and rest it on the mouth of the jar. Put three or four ice cubes inside the lid.

3. **Observe** Watch the underside of the lid for ten minutes.

4. **Communicate** Record your observations.

_____

_____

Unit D · Water and Weather   Use with textbook page D77   135

©Macmillan/McGraw-Hill

Name_____ Date_____

Drawing Conclusions

1. **Infer** Where did the moisture on the underside of the lid come from?
   _____
   _____

2. **Infer** Why was it necessary to cool the lid with ice to make "raindrops"?
   _____
   _____

3. What evaporated in this experiment? What condensed?
   _____
   _____
   _____

4. **FURTHER INQUIRY** **Experiment** How does the temperature of the water in the jar affect the model of rain? Predict first and then set up an experiment to answer the question.
   _____
   _____

Inquiry

Think of your own questions that you might like to test. How can different temperatures affect the water cycle?

My Question Is:
_____

How I Can Test It:
_____

My Results Are:
_____
_____

136 | Unit D · Water and Weather | Use with textbook page D77

Name_____ **Date**_____

Alternative Explore
Lesson 7

Making It Rain in a Bowl

Procedure

1. Watch your teacher pour a cup of hot water into a clear bowl and then cover it with plastic wrap.

2. Observe the underside of the plastic wrap for a few minutes. Draw what you see below.

Materials

- clear bowl
- cup of hot water
- plastic wrap

Drawing Conclusions

1. Where did the moisture on the underside of the plastic wrap come from?

_____

_____

2. What caused the moisture to collect on the plastic wrap?

_____

_____

_____

Unit D · Water and Weather   Use with TE textbook page D77

137

Name_____ Date_____

Explore Activity
Lesson 1

How Can You Identify Matter?

Hypothesize Is air matter? Use what you know about matter to find out.

Write a **Hypothesis:**

_____

_____

Materials

- 2 identical balloons
- meterstick
- string
- scissors
- tape

Procedure: Design Your Own

BE CAREFUL! Handle scissors carefully.

1. **Experiment** Using the materials, design an experiment to determine whether air is matter.

_____

_____

_____

_____

_____

2. **Use Variables** Do the experiment. Record each step, all your observations, and your results.

_____

_____

3. Repeat each step of your experiment three times. Record any differences in the results of each trial.

_____

Drawing Conclusions

1. Is air matter? What evidence do you have to support your conclusion?

_____

_____

138   Unit E · Matter   Use with textbook page E5

Name_____ Date_____

2. Use evidence to explain why you may have gotten different results with some trials. How does this affect your conclusion?

_____
_____

3. **FURTHER INQUIRY** Infer Think of another object that is filled with air. Does it also provide evidence that air is matter? Explain.

_____
_____
_____

Inquiry

Think of your own questions that you might like to test. What are the effects of moving air, such as the wind, on objects?

My Question Is:

_____
_____

How I Can Test It:

_____
_____

My Results Are:

_____
_____

Unit E · Matter   Use with textbook page E5

Name_____ Date_____

Blowing Bubbles

Alternative Explore
Lesson 1

Materials
- bottle of bubble solution
- wand

Procedure

1. With a partner, take turns blowing bubbles.

2. What happens when the bubbles are first blown from the wand?
   _____
   _____

3. What happens to the bubbles after a while?
   _____
   _____

Drawing Conclusions

1. How do you know that air was inside the bubbles?
   _____
   _____

2. How do you know that air is matter?
   _____
   _____
   _____

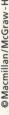

140   Unit E · Matter   Use with TE textbook page E5

Name_____ **Date**_____

QUICK LAB
FOR SCHOOL OR HOME
Lesson 1

Is Bigger Always More?

Hypothesize Does a larger object always have more mass than a smaller object? Write a **Hypothesis:**

_____

_____

Materials
- meterstick
- balloon
- quarter
- string
- scissors
- tape

Procedure

1. **Predict** Which has more mass, an inflated balloon or a quarter? Record your prediction.

   _____

   _____

2. **Experiment** Inflate a balloon. Attach it to one end of a meterstick. Attach a quarter to the other end.

3. **Observe** What happens to the meterstick? Record your observations.

   _____

   _____

©Macmillan/McGraw-Hill

Unit E · Matter   Use with textbook page E10   141

Name_____ Date_____

QUICK LAB
FOR SCHOOL OR HOME
Lesson 1

Drawing Conclusions

4. Which object has more mass?

_____

5. Infer Do small objects ever have more mass than larger objects?

_____

6. Name some examples of matter that are smaller than a balloon but have more mass than a balloon.

_____

_____

7. Going Further The balloon filled with air has some mass but not as much as the other objects to which you compared it. Can objects be the same size but have different masses?

My Hypothesis Is:

_____

_____

My Experiment Is:

_____

_____

_____

My Results Are:

_____

_____

_____

142   Unit E · Matter   Use with textbook page E10

Name_____ Date_____

How Can You Measure Matter?

Hypothesize What methods can you use to compare sizes of objects with different shapes? Write a **Hypothesis**:
_____
_____

Materials
- 3 different paper shapes labeled A, B, and C
- ruler
- pencil

Procedure

1. **Observe** Carefully look at shapes A and B.

2. **Predict** Predict which shape you think is bigger. Record your prediction.
_____
_____

3. **Measure** Use the ruler to draw one-inch squares on each shape. How can you use the squares to determine which shape is bigger?
_____

4. Follow steps 2 and 3 to compare shapes A and B with shape C. Put all three in order from smallest to largest.
_____

Drawing Conclusions

1. **Communicate** How did you compare the shapes in step 2? In step 3?
_____
_____

2. **Infer** Which method is more accurate? Why?
_____
_____

Unit E · Matter   Use with textbook page E15   143

Name_____ **Date**_____

Explore Activity
Lesson 2

3. **Classify** Which paper was the largest? Smallest?

_____

4. **FURTHER INQUIRY** **Experiment** Make two different shapes of your own. Challenge a classmate to determine which is larger.

Inquiry

Think of your own questions that you might like to test. What other ways of comparing shapes are there?

My Question Is:

_____

_____

How I Can Test It:

_____

_____

_____

My Results Are:

_____

_____

©Macmillan/McGraw-Hill

144   Unit E · Matter   Use with textbook page E15

| Name_____ | Date_____ |

Alternative Explore
Lesson 2

Shopping for Standard Measurements

Procedure

1. Imagine that you are shopping for products in a supermarket where no standard measurements are used.

2. Think of things buyers and sellers might disagree about and ways they might reach agreement.

Drawing Conclusions

1. What are some products that you need standard measurements to buy?

_____

_____

_____

2. What disagreements might a buyer and seller have about a product if there were no standard measurements used?

_____

_____

_____

3. How could the buyer and seller reach agreement without using standard measurements?

_____

_____

_____

©Macmillan/McGraw-Hill

Unit E · Matter   Use with TE textbook page E15   145

Name	Date

Inquiry Skill Builder
Lesson 2

Infer

Examine If Shape Affects Volume

Does the volume of an object change if its shape changes? In this activity you will use water to help you find the volume of clay molded into different shapes. You will use your observations and measurements to infer the answer to the question. When you infer, you use observations to figure something out.

Materials

- clay
- graduated cylinder
- water
- string
- paper towels

Procedure

1. Fill the graduated cylinder with 50 mL of water.

2. Make a solid figure out of the clay. Press the string into it.

3. **Observe** Hold the string. Lower the clay into the water until it is completely covered. Carefully observe and graph the new water level. Remove the figure.

_____

_____

4. Repeat step 3 two times to verify your results.

5. Rearrange the clay to make a different shape. Do not add or take away any clay.

6. **Predict** if the new shape will have the same volume as the first. Repeat step 3 with the new shape. Repeat two more times to verify your results.

_____

_____

Drawing Conclusions

1. **Measure** What was the volume of each figure? How did you find out?

_____

_____

146   Unit E · Matter   Use with textbook page E18

Name _____ **Date** _____

Inquiry Skill Builder
Lesson 2

2. Infer Does an object's volume change when you change its shape? How do you know?

_____

_____

3. Infer Toy A raised the water level in a tank 1 cm. Toy B raised the water level 2 cm. What can you infer about their volumes?

_____

_____

_____

Unit E · Matter   Use with textbook page E18

Name_____ Date_____

QUICK LAB
FOR SCHOOL OR HOME
Lesson 2

Comparing Densities

Hypothesize How can you compare densities of different items? Write a **Hypothesis:**

_____

_____

_____

Materials

- equal-sized samples of a wooden block, clay, and foam
- pan balance
- metric ruler

Procedure

1. Obtain samples from your teacher.
   Does each sample have the same volume?
   Sketch each mineral.

   _____

   _____

2. **Predict** Which sample do you think has the greatest density?
   The least? Write your predictions.

   _____

   _____

3. **Measure** Use the balance to compare the masses of the samples.
   Record the data in the table below.

Material	Mass
wood	
clay	
foam	

© Macmillan/McGraw-Hill

148   Unit E · Matter   Use with textbook page E20

Name_____ Date_____

Drawing Conclusions

4. Rank the items from greatest to least density.
_____
_____

5. **Infer** Why would you need information about both mass and volume to compare density?
_____
_____

6. **Going Further** Does the density of a material change when it is made into a different shape? Write down your thoughts. How can you verify your ideas using the clay sample?

My Hypothesis Is:
_____
_____

My Experiment is:
_____
_____
_____

My Results Are:
_____
_____
_____

Unit E · Matter   Use with textbook page E20   149

Name_____ Date_____

Explore Activity
Lesson 3

How Can You Classify Matter?

Hypothesize How would you design a classification system for ten items? How can you test your ideas? Write a **Hypothesis**:

_____
_____
_____

Materials
- 10 assorted items
- index cards

Procedure: Design Your Own

1. Write the name of each item on an index card.

2. **Classify** Sort the items into groups based on properties you can observe. Record the properties. Use a system like the one shown here. Then try another system.

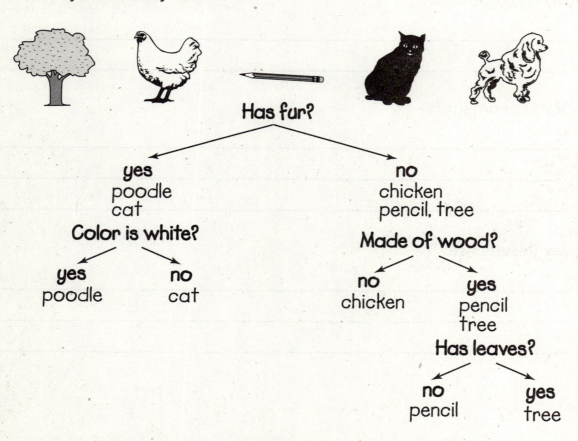

_____
_____

150   Unit E · Matter   Use with textbook page E31

Name_____ Date_____

Explore Activity
Lesson 3

Drawing Conclusions

1. What properties did you use to classify the items?
   _____
   _____

2. **Communicate** Were you able to place all the items into the groups? Why or why not? Write a paragraph explaining your answer.
   _____
   _____
   _____
   _____
   _____

3. **FURTHER INQUIRY** **Classify** Trade cards with a classmate to see if he or she can follow your classification system. Have the classmate identify the object that has been classified.
   _____

Inquiry

Think of your own questions that you might like to test. What other things would you like to classify?

My Question Is:
_____

How I Can Test It:
_____

My Results Are:
_____
_____
_____

Unit E · Matter   Use with textbook page E31   151

Name_____ **Date**_____

Alternative
Explore
Lesson 3

Parts of Speech

Procedure

1. Your teacher will give you a list of words.

2. You will work with your group to classify the words. First, as a group, decide what the basis for classifying the words will be.

3. In the space below, make a diagram to show how you classified your words into groups.

Drawing Conclusions

1. On what basis did your group classify the words?

_____

_____

2. Why is it useful to classify things?

_____

_____

©Macmillan/McGraw - Hill

152   Unit E · Matter   Use with TE textbook page E31

Name_____ **Date**_____

QUICK LAB
FOR SCHOOL OR HOME
Lesson 3

Mix and Unmix

Hypothesize How can you use physical properties to separate the parts of a mixture? Write a **Hypothesis:**

_____

_____

_____

Procedure

BE CAREFUL! Wear goggles.

1. **Observe** Design an experiment to separate a mixture. You may use more than one method. Record the methods you used as well as your observations and results.

_____

_____

_____

_____

_____

_____

Drawing Conclusions

2. **Observe** Were you able to completely separate the mixture? How do you know?

_____

_____

_____

Materials

- goggles
- mixture from your teacher
- piece of dark paper
- magnet
- filter paper
- plastic funnel
- tweezers or forceps
- container
- water

©Macmillan/McGraw-Hill

Unit E · Matter

Use with textbook page E36

153

Name _____ **Date** _____

QUICK LAB
FOR SCHOOL OR HOME
Lesson 3

3. Form a Hypothesis How could you separate a mixture of white sand and salt? Test your ideas.

_____

_____

_____

4. Going Further Sometimes it is not obvious that something is a mixture. Heavy cream contains butterfat and water. Can you separate the water from the butterfat in heavy cream? Write down your idea and try it. The separation need not be complete.

My Hypothesis Is:

_____

_____

My Experiment Is:

_____

_____

_____

My Results Are:

_____

_____

©Macmillan/McGraw-Hill

154   Unit E · Matter   Use with textbook page E36

Name_____ Date_____

Explore Activity
Lesson 4

How Can Things Change?

Hypothesize How do changes you make to an object affect the matter the object is made of? Write a Hypothesis:

_____

_____

_____

Procedure

BE CAREFUL! Handle scissors and a knife carefully.

1. Record each object in the first column of the table below.

2. **Observe** Carefully observe each object. What properties does each have? Record the properties in the second column.

3. **Predict** How can you change each object? Record your predictions in the third column.

4. **Experiment** Test your predictions. Observe each object after the change. What properties does each have now? Record the properties in the last column.

Materials

- flat sheet of paper
- modeling clay
- scissors
- rubbing alcohol
- ice cube
- plastic plate with rim
- whole fruit, such as an orange, lemon, or apple
- plastic knife

Object	Properties	Changes	Properties After Changes

©Macmillan/McGraw-Hill

Unit E · Matter   Use with textbook page E43   155

Name_____ Date_____

Drawing Conclusions

1. **Communicate** How are the properties of each object similar and different before and after the changes?
   _____
   _____

2. **Infer** Do you think the changes you made actually changed the matter making up each object? Why or why not?
   _____
   _____

3. FURTHER INQUIRY **Communicate** Make more changes to different types of matter. Use your observations to explain that things change in some ways and stay the same in other ways.
   _____
   _____

Inquiry

Think of your own questions that you might like to test. What physical changes do we see around us every day?

My Question Is:
_____
_____

How I Can Test It:
_____
_____
_____

My Results Are:
_____
_____

Unit E · Matter   Use with textbook page E43

Name_____ Date_____

Identify Changes

Procedure

BE CAREFUL! Use scissors carefully.

Materials
- paper samples
- scissors

1. List some properties of the paper samples.
   _____
   _____
   _____

2. Use the scissors to change the paper samples. What kinds of changes did you make?
   _____
   _____
   _____

Drawing Conclusions

1. What other ways could you make changes to the paper?
   _____
   _____
   _____

2. Did the changes you make actually change the matter making up each paper sample?
   _____
   _____
   _____

Unit E · Matter   Use with TE textbook page E43

| Name_____ | Date_____ |

Inquiry Skill Builder
Lesson 4

Experiment

How Heat Energy Affects Evaporation

When you perform an experiment, you first form a hypothesis. Then you test your hypothesis. Follow the steps to test how heat energy affects evaporation.

Procedure

1. Place a wet paper towel across the top of each plastic glass. Secure each with a rubber band.

2. **Hypothesize** Place one glass where you think the paper towel will dry fastest. Place another where you think it will dry slower. Place the third where you think it will dry slowest.

_____

_____

_____

3. **Measure** Use the thermometer to measure the temperature near each glass. Record the temperatures.

_____

_____

_____

4. Record the time you start timing. Then touch each paper towel every two minutes. Record the time the first paper towel is dry.

_____

_____

_____

5. Repeat step 4 until the other towels are dry.

Materials

- 3 paper towels
- 3 rubber bands
- three 10-oz clear-plastic glasses
- thermometer
- clock or watch
- desk lamp
- container of water

©Macmillan/McGraw-Hill

158   Unit E · Matter   Use with textbook page E48

Name_____ **Date**_____

Inquiry Skill Builder
Lesson 4

Drawing Conclusions

1. **Interpret** In which place did a paper towel dry the fastest? What was the temperature?

   _____

   _____

2. **Interpret** In which place did a paper towel dry the slowest? What was the temperature?

   _____

   _____

3. **Experiment** Would water evaporate from a paper towel as fast if you put an inverted glass over it? Try it.

   _____

   _____

   _____

   _____

©Macmillan/McGraw-Hill

Unit E · Matter   Use with textbook page E48   159

Name_____ Date_____

What Causes the Change?

Hypothesize Compare the Statue of Liberty with clean copper. What do you think caused the statue to turn green? Write a **Hypothesis:**

_____
_____
_____

Procedure

BE CAREFUL! Wear goggles.

1. Put a small wad of modeling clay on the bottom of the petri dish or plastic glass.

2. Wedge the penny in the clay so that it is vertical.

3. Add vinegar to cover the bottom of the petri dish or glass. Cover the petri dish with the plastic glass. If you put the penny in the glass, cover the glass tightly with plastic wrap.

4. **Predict** What do you think will happen to the penny? Record your prediction.

_____
_____

Materials
- goggles
- petri dish or 10-oz clear-plastic glass
- modeling clay
- shiny penny
- vinegar
- 10-oz clear-plastic glass or plastic wrap

Drawing Conclusions

1. **Observe** What happens to the penny after one hour? After three hours? Overnight? Record your observations.

_____

160   Unit E · Matter   Use with textbook page E51

Name_____ Date_____

2. How is this penny different from the penny that your teacher soaked in vinegar overnight?
_____

3. Form a Hypothesis What do you think caused the changes to your penny but not the soaked penny?
_____
_____
_____

4. FURTHER INQUIRY **Experiment** Do you think other materials would change also? Repeat the activity using a paper clip.
_____

Inquiry

Think of your own questions that you might like to test. What other substances might cause similar reactions?

My Question Is:
_____

How I Can Test It:
_____
_____

My Results Are:
_____
_____

Unit E · Matter   Use with textbook page E51

Name_____ Date_____

Alternative Explore
Lesson 5

Other Changes

Procedure

BE CAREFUL! Wear goggles to protect your eyes.

1. Put a small wad of modeling clay on the bottom of the plastic cup.

2. Wedge the penny in the clay so that it is vertical.

3. Choose a liquid to test. Pour 1 tsp. of the liquid into the cup. Cover the cup tightly with plastic wrap.

4. Predict what will happen to the penny.

_____

_____

5. Observe the penny every day for a few days. Record your observations.

_____

_____

_____

Materials

- goggles
- 10-oz clear plastic cup
- modeling clay
- shiny penny
- 1 tsp. milk or other liquid
- 10-oz clear-plastic glass or plastic wrap

Drawing Conclusions

1. Did you observe a change in the penny?

_____

2. How did your observations compare to your prediction?

_____

162   Unit E · Matter   Use with TE textbook page E51

©Macmillan/McGraw-Hill

Name_____ **Date**_____

QUICK LAB
FOR SCHOOL OR HOME
Lesson 5

Preventing Chemical Change

Hypothesize Why do you think most pennies you use every day aren't green? Write a **Hypothesis:**

_____

_____

_____

_____

Materials

- goggles
- petri dish or 10-oz clear-plastic glass
- modeling clay
- 3 shiny pennies
- 1 tsp. vinegar
- 10-oz clear-plastic glass or plastic wrap
- other materials as needed

Procedure

BE CAREFUL! Wear goggles.

1. **Hypothesize** Follow the procedure in the Explore Activity. Think about something you can do to keep the penny from turning green. Record your ideas.

_____

_____

2. **Experiment** Test your ideas. Record your results.

_____

_____

_____

Drawing Conclusions

3. **Compare** Make a class table of the results for each test. What kept the pennies from turning green?

_____

_____

_____

_____

© Macmillan/McGraw-Hill

Unit E · Matter   Use with textbook page E55   **163**

Name_____ **Date**_____

QUICK LAB
FOR SCHOOL OR HOME
Lesson 5

4. **Infer** What do you think prevents the pennies you use every day from turning green?

_____

_____

5. **Going Further** Is there more than one reason why the pennies we use every day are not green? Is the environment where you live very damaging to pennies? Is the coating that forms on everyday pennies protective?

My Hypothesis Is:

_____

_____

My Experiment Is:

_____

_____

_____

_____

My Results Are:

_____

_____

164   **Unit E · Matter**   **Use with textbook page E55**

©Macmillan/McGraw-Hill

Name_____ Date_____

Explore Activity
Lesson 1

How Can You Tell Something Is Moving?

Hypothesize What does the distance traveled tell about the speed of movement?

Write a **Hypothesis:**

_____

_____

Materials
- paper
- marking pens
- watch or clock

Procedure

1. Make a map of your classroom. Mark where the main objects are placed. For example, show the location of doors, windows, chalkboards, and your desks.

2. For five minutes, slowly follow a path around the room. Stop once every minute, and mark an X on the map to show your location. Mark the time. Rest at some places for a short time. Next to those Xs, write the time that you got there and the time you left.

3. **Communicate** Trade maps with your partner. Determine at which positions he or she was moving.

_____

_____

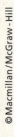

Unit F · Energy   Use with textbook page F5   165

Name_____ Date_____

Drawing Conclusions

1. **Interpret Data** How could you tell when your partner was moving? Standing still?
   _____
   _____

2. **Infer** Who is moving faster—a student whose Xs are far apart or one whose Xs are close together? Give the reason for your answer.
   _____
   _____

3. FURTHER INQUIRY **Use Variables** Repeat the activity by making a map of your home. Is your motion visible?

Inquiry

What else could you test for movement? Think about an object that can move. How can you tell whether or not it has moved?

My Question Is:
_____
_____

How I Can Test It:
_____
_____

My Results Are:
_____
_____

166   Unit F · Energy   Use with textbook page F5

Name_____ **Date**_____

Alternative Explore
Lesson 1

The Marble Path

Procedure

1. Work in groups of three. Choose a room at home or in school with a smooth floor. One group member places a marble on the floor at some point and taps it gently to make it move.

2. Another member times the marble with a watch until it stops moving.

3. A third member follows the marble carefully as it moves. Use any surrounding objects (chairs, tables) to help you remember its path. When the marble stops, lay a string along the floor to trace the path.

4. Measure the length of the path.

   _____

5. Repeat several times, switching roles.

Materials

- marbles
- stopwatch or clock with second hand
- string
- meterstick

Student Name	Marble Time	Distance

Drawing Conclusions

Which time did the marble moved fastest? How can you tell?

_____

_____

_____

© Macmillan/McGraw-Hill

Unit F · Energy   Use with TE textbook page F5   167

Name_____ Date_____

Inquiry Skill Builder
Lesson 1

Measure

Marble Motion

How fast can a marble move? You could answer this question by saying "fast" or "slow." However, if you can make measurements, you can determine speed more exactly.

Procedure

1. **Measure** Work with two partners. Measure the length of the tube with a ruler. Record your measurement.

   _____

2. Stack three books to make a tower. Place one end of the tube on the books. Let the other end of the tube touch the fourth book, which is standing straight up. Tape the tube in place.

3. **Measure** Time the marble as it rolls down the tube. One partner should release the marble, while another partner starts the stopwatch. The third partner should hold the standing book in place. When you hear the marble hit the book, record the time.

   _____

4. **Measure** Repeat step 3 ten times. Calculate the speed for each trial.

   _____

   _____

   _____

   _____

Materials

- long card-board tube or several shorter tubes taped together
- marble
- 4 books
- ruler
- tape
- stopwatch

168   Unit F · Energy   Use with textbook page F9

Name_____ Date_____

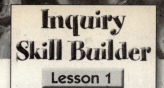

Drawing Conclusions

1. What two measurements did you need to calculate speed?
   _____

2. **Use numbers** What equation did you use to calculate speed?
   _____

3. Remove one book to change the height of the ramp. How does this change the speed of the marble?
   _____
   _____

Name_____ Date_____

Explore Activity
Lesson 2

How Does a Pulley Work?

Hypothesize How could the movers get a chest to the third floor? Write a **Hypothesis**:

_____

Materials

- pulley
- 2 pieces of rope (thick string)
- book
- spring scale

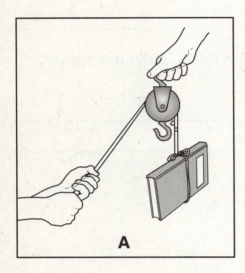
A

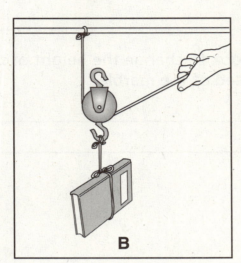
B

Procedure

1. Tie the longer piece of cord around the book. Have a partner hold the pulley. Thread the cord through the pulley's groove.

2. **Observe** Pull down on the cord. What happens? Draw the pulley, cord, and book system. Use arrows to show the direction you pull and the direction the book moves.

_____

3. **Observe** Attach one end of the second cord to something that won't move. Thread the free end through the pulley's groove so that you can pull up on it. Attach the book to the hook on the pulley. Pull up on the other end of the cord. What happens to the pulley? What happens to the book? Draw the system. Use arrows to show the direction you pull and the direction the book and pulley move.

_____

170 Unit F · Energy Use with textbook page F19

Name_____ Date_____

Drawing Conclusions

1. Was it easier to lift the book in step 2 or 3? Why do you think so?

   _____
   _____

2. **Infer** In which step did you pull one way and the book moved the opposite way?

   _____

3. FURTHER INQUIRY **Experiment** Repeat the procedures but attach a spring scale to the end of the cord you pull on. Record the readings on the scale for each trial. What do you notice?

   _____
   _____
   _____

Inquiry

What other loads would you like to test with the pulley systems? Write and conduct an experiment.

My Question Is:

_____
_____

How I Can Test It:

_____
_____

My Results Are:

_____
_____
_____

Unit F · Energy   Use with textbook page F19   171

Name_____ Date_____

Alternative Explore
Lesson 2

Pulley Systems

Procedure

Materials
- 2 or 3 pulleys
- several pieces of rope
- book
- spring scale

1. You will work in a group to design a pulley system to make the job of lifting a book easier.

2. Open the book and place a rope in it. Close the book and tie the ends of the rope together so that the book may be lifted by the rope.

3. Use the spring scale to find the amount of force needed to lift the book.

4. Assemble a pulley system using two or more pulleys.

5. Use the spring scale to find out the amount of force needed to lift the book using the pulley system. Draw your pulley system.

Drawing Conclusions

1. Did the pulley system change the amount of force needed to lift the book? If so, by how much?
_____
_____

2. How do you think the amount of force needed to lift the book would change if you could add more pulleys to your system?
_____
_____

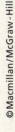

172 Unit F · Energy Use with TE textbook page F19

Name_____ Date_____

Make Levers

Hypothesize What happens to the direction of the force when you use a lever? Write a **Hypothesis**:
_____
_____
_____

Materials
- book
- ruler
- pencil

Procedure

1. **Make a Model** Place about an inch of a ruler under the edge of a book. Place a pencil under the ruler close to the book.
   _____
   _____

2. Push down on the other end of the ruler. Record what happens.
   _____

3. Place as much of the ruler under the book as fits. Remove the pencil.

4. Lift up on the end of the ruler sticking out from under the book. Record what happens.
   _____

Drawing Conclusions

5. **Draw Conclusions** Can a lever change the direction of the force? Explain.
   _____

6. **Draw Conclusions** What kind of lever did you make? Explain.
   _____
   _____

Unit F · Energy — Use with textbook page F23 — 173

Name_____ **Date**_____

QUICK LAB
FOR SCHOOL OR HOME
Lesson 2

7. **Going Further** Demonstrate the mechanical advantage of a second-class lever. Where is the load located when the effort force is the smallest? Where is the load located when the effort force is the greatest? Write and conduct an experiment.

My Hypothesis Is:

_____

_____

My Experiment Is:

_____

_____

_____

_____

_____

_____

_____

_____

_____

My Results Are:

_____

_____

174   Unit F · Energy   Use with textbook page F23

©Macmillan/McGraw-Hill

Name _____ **Date** _____

Explore Activity
Lesson 3

How Does Fat Keep Mammals Warm?

Hypothesize Walruses have a thick layer of body fat beneath their skin. Do you think it keeps them warm in the freezing waters?

Write a **Hypothesis:**

_____

_____

Materials

- plastic bag containing lard or solid vegetable shortening

- 2 vinyl surgical gloves

- bucket or pan of ice water

- stopwatch or clock with second hand

- paper towels

Procedure

1. **Predict** Do you think an extra layer of fat can help keep your hand warm in very cold water? Why or why not?

   _____

   _____

2. **Measure** Put on one glove. Ask your partner to time how long you can comfortably keep your hand in the ice water. Record the results.

   _____

3. **Use Variables** Move your gloved hand around in the bag of lard to coat it well. Be sure to spread the lard over your entire hand and between your fingers.

4. **Measure** Ask your partner to time how long you can comfortably keep your lard-coated hand in the ice water. Record the results. Trade places, and let your partner repeat the procedure.

   _____

5. **Use Numbers** Take an average of both of your results.

   _____

© Macmillan/McGraw-Hill

Unit F · Energy | Use with textbook page F33 | 175

Name_____ Date_____

Drawing Conclusions

1. **Communicate** How long on average were you able to keep your hand in the ice water in step 2? In step 4?
   _____

2. **Infer** If the lard represents a walrus's blubber, how might blubber help the walrus survive?
   _____
   _____

3. FURTHER INQUIRY **Infer** Repeat at different temperatures. Collect data and graph the results. How does the temperature affect the time in the water?
   _____
   _____

Inquiry

Think of your own questions that you might like to test. What other forms of insulation would you like to test?

My Question Is:
_____
_____

How I Can Test It:
_____
_____

My Results Are:
_____
_____

176   Unit F · Energy   Use with textbook page F33

Name_____ **Date**_____

Alternative Explore
Lesson 3

Fatten Up

Procedure

1. Put on the gloves.

2. Move your gloved hands around in the bag of lard to coat it well. Be sure to spread the lard over your entire hand and between your fingers. Make sure that one hand is more thickly coated than the other.

3. Ask your partner to time you while you put your lard-coated hands in the ice water. When either hand becomes uncomfortable, remove it and ask your partner to note the time.

Drawing Conclusions

1. Which hand were you able to keep in the cold water longer? Explain your answer.

_____

_____

2. How do your results compare to the way winter jackets are made?

_____

_____

Materials

- plastic bag containing lard or solid vegetable shortening

- 2 vinyl surgical gloves

- bucket or pan of ice water

- timer or watch with second hand

- paper towels

©Macmillan/McGraw-Hill

Unit F · Energy | Use with TE textbook page F33 | 177

Name_____ Date_____

QUICK LAB
FOR SCHOOL OR HOME
Lesson 3

Matter and Heat

Hypothesize What happens to air when it is heated?
Write a **Hypothesis:**

_____
_____
_____

Materials

- goggles
- inflated balloon
- blow dryer
- string
- ruler
- marking pen
- timer or watch with second hand

Procedure

BE CAREFUL! Wear goggles.

1. **Measure** Use a string and ruler to measure a blown-up balloon. Mark the spot where you measured it. Record the data.

_____

_____

2. **Measure** Heat the balloon with a blow dryer for one minute. Measure the distance around the balloon at the marked spot. Record your observations and measurements.

_____

Drawing Conclusions

3. How did heat affect the size of the balloon?

_____

_____

4. **Communicate** What happens to the air particles in the balloon when the balloon is heated?

_____

_____

_____

©Macmillan/McGraw-Hill

178   Unit F · Energy   Use with textbook page F37

Name _____ **Date** _____

QUICK LAB
FOR SCHOOL OR HOME
Lesson 3

5. **Going Further** Can you demonstrate a change in volume with temperature using the balloon and a different type of heat transfer? Write and conduct an experiment.

My Hypothesis Is:

_____

_____

My Experiment Is:

_____

_____

_____

My Results Are:

_____

_____

_____

©Macmillan/McGraw-Hill

Unit F · Energy   Use with textbook page F37   179

Name_____ Date_____

Explore Activity
Lesson 4

What Do You See When You Mix Colors of Light?

Hypothesize How can you show that white light is made up of all the colors?

Write a **Hypothesis:**

_____

_____

Materials
- cardboard
- compass
- scissors
- markers or colored crayons
- pencil
- goggles

Procedure

BE CAREFUL! Wear goggles.

1. **Measure** Use a compass to draw a circle on the cardboard. The circle should be about 13 cm (5 in.) across. Divide the circle into 12 equal sections. Color each section a different color.

2. Cut out the circle. Put on your goggles. Carefully push a pencil into the center of the circle. Spin your spinner away from your body.

3. **Observe** What color do you see while the spinner is spinning?

_____

4. **Experiment** Repeat steps 1–3 to make another spinner. This time choose colors that you think will make the disk appear white when you spin it. Make as many spinners as you need to find the color combinations that work best.

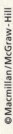
©Macmillan/McGraw-Hill

180   Unit F · Energy   Use with textbook page F41

Name_____ Date_____

Explore Activity
Lesson 4

Drawing Conclusions

1. What color did you see in step 3?
   _____

2. What colors did you mix together to make the best white?
   _____

3. **FURTHER INQUIRY** **Experiment** Repeat steps 1–3 to make another spinner. This time color the spinner only three colors, and see if you can make the spinner still look white. Make as many spinners as you need to find the color combinations that work best.

Inquiry

What colors could you mix together to make colors besides white? Write and conduct an experiment.

My Question Is:
_____
_____

How I Can Test It:
_____
_____

My Results Are:
_____
_____

Unit F · Energy   Use with textbook page F41   **181**

Name_____ Date_____

Seeing Different Colors

Procedure

1. Shine your flashlight through the prism. Try holding the prism at different angles. What colors do you see?
   _____
   _____

Materials
- prism
- flashlight

2. How does the prism appear when you look at it without the flashlight?
   _____
   _____

Drawing Conclusions

1. What can cause light to break into its colors?
   _____
   _____

2. You can see the colors of light in a rainbow. Where else have you seen these colors?
   _____
   _____

Name_____ Date_____

QUICK LAB
FOR SCHOOL OR HOME
Lesson 4

Absorbing Light

Hypothesize Do dark or light colors absorb more heat?
Write a **Hypothesis:**

_____

_____

Materials

- 2 thermometers
- black paper
- white paper
- sunny windowsill or desk lamp
- clock

Procedure

1. **Predict** Wrap a thermometer in black paper. Wrap a second thermometer in white paper. Put the two thermometers on a sunny windowsill or under a desk lamp. Predict which thermometer will heat up faster. Write your prediction.

   _____

   _____

2. **Measure** Read and record each temperature after ten minutes.

   _____

Drawing Conclusions

3. Which thermometer warmed up faster?

   _____

4. **Infer** Why do you think this happened?

   _____

5. **Hypothesize** What do you think would happen if you tried different colors of paper? Test your hypothesis.

   _____

   _____

©Macmillan/McGraw-Hill

Unit F · Energy   Use with textbook page F49   **183**

Name_____ **Date**_____

QUICK LAB
FOR SCHOOL OR HOME
Lesson 4

6. **Going Further** Do darker colors of other materials absorb more light energy than light colors? Write and conduct an experiment.

My Hypothesis Is:

_____

_____

My Experiment Is:

_____

_____

_____

_____

My Results Are:

_____

_____

184   Unit F · Energy   Use with textbook page F49

Name_____ Date_____

How Do Vibrations Produce Sound?

Hypothesize How are sounds made when a guitar is played? Write a **Hypothesis:**

_____

_____

Materials
- milk carton
- fishing line or string
- wire
- rubber band
- scissors

Procedure

BE CAREFUL! Handle scissors carefully.

1. Cut off the top and one of the sides of the milk carton as shown in the picture in your textbook. Make a hole in the bottom of the carton.

2. Thread the string through the hole. Tie a knot at one end to keep it from slipping through the hole.

3. **Observe** Hold the box at the end near the hole. Ask your partner to hold the free end of the string and pluck it. Record what you hear. What happens to the string?

_____

_____

4. **Experiment** Repeat the activity with wire and then with a rubber band.

Unit F · Energy   Use with textbook page F53   185

Name_____	Date_____

Explore Activity
Lesson 5

Drawing Conclusions

1. **Communicate** How did you make sound?

   _____

2. **Compare** How does the sound of the wire differ from the sound of the string? The rubber band?

   _____

3. **FURTHER INQUIRY** **Use Variables** Try changing the length of the string. What happens to the sound?

   _____

Inquiry

Think of your own questions that you might like to test about sound. How do other musical instruments make sound?

My Question Is:

_____

_____

How I Can Test It:

_____

_____

My Results Are:

_____

_____

©Macmillan/McGraw-Hill

Name_____ **Date**_____

Alternative Explore
Lesson 5

Measuring Sound

Procedure

1. Hold one end of a ruler tightly against the top of your desk. Let most of the ruler hang over the edge of the desk.

2. Gently pull up the free end of the ruler. Then let go of it.

3. Repeat steps 1 and 2 with each of the different rulers. Place the rulers in sound order. The one that makes the highest sound should come first. The one that makes the lowest sound should come last.

_____

_____

_____

_____

4. Experiment with each ruler. Make the part that moves shorter than before. How did the sounds change?

_____

Drawing Conclusions

1. What is the source of the sounds?

_____

_____

2. What makes a sound high or low?

_____

_____

3. Name some musical instruments that make movements you can see as they make sounds.

_____

Materials

- several thin metal, plastic, and/or wood rulers

© Macmillan/McGraw-Hill

Unit F · Energy

Use with TE textbook page F53

187

Name_____ Date_____

Explore Activity
Lesson 6

What Happens to Rubbed Balloons?

Materials

- two 9-in.-round balloons, inflated
- 2 pieces of string, 50 cm each
- tape
- wool cloth

Hypothesize What do you think will happen when two rubbed balloons are brought next to each other? Will they pull together or push apart?

Write a **Hypothesis:**

_____

_____

Procedure

1. Tie a piece of string to each inflated balloon. Hang them from a table as shown. Tape the loose end of the string to the table.

2. Observe any action of the balloons.

_____

3. **Predict** What do you think will happen if you rub one balloon with a piece of wool cloth? Both balloons? Record your predictions.

_____

4. Experiment. Test your predictions.

_____

5. **Predict** What do you think will happen if you hold the wool cloth between the balloons? Test your prediction.

_____

_____

6. **Predict** What do you think will happen if you put your hand between the two balloons? Test your prediction.

_____

©Macmillan/McGraw-Hill

188   Unit F · Energy   Use with textbook page F69

Name_____ Date_____

Drawing Conclusions

1. **Communicate** What happened when you rubbed one balloon with the wool cloth? Both balloons?
_____
_____

2. **Communicate** What happened when you put the wool cloth between the balloons? Your hand?
_____
_____
_____

3. FURTHER INQUIRY **Experiment** Untie one balloon. Rub it with the wool. Try to stick it to the wall. What happens? Why?
_____
_____
_____

Inquiry

Think of your own questions that you might like to test using the balloons and wool cloth. What will happen if the balloons are rubbed with the cloth more than once?

My Question Is:
_____
_____

How I Can Test It:
_____
_____
_____

My Results Are:
_____
_____

Unit F · Energy   Use with textbook page F69

Name_____ Date_____

Alternative Explore
Lesson 6

Charging Hands

Procedure

1. Tie one end of the string to the balloon. Tape the other end of the string to the edge of a table. The balloon should hang free and not touch anything.

2. Wash and dry your hands well.

3. Bring your hand near the balloon, but do not touch it. Record what happens.

_____

4. Hold the balloon with one hand and rub it briskly 10 times with the palm of your other hand. Release the balloon.

5. Then bring the hand you rubbed the balloon with near the balloon, but do not touch it. Describe what happens.

_____

_____

Materials

- inflated balloon

- 12-inch length of string

- tape

Drawing Conclusions

1. What difference did you see between the results of steps 3 and 5?

_____

_____

2. Explain why this happened.

_____

_____

_____

_____

Unit F · Energy   Use with TE textbook page F69

Name_____ Date_____

Making Static Electricity

Quick Lab — FOR SCHOOL OR HOME — Lesson 6

Hypothesize How can you make static electricity? How will the static electricity affect objects?

Write a **Hypothesis:**
_____
_____

Materials
- plastic comb
- mirror
- faucet with running water
- inflated balloon

Procedure

1. Comb your hair several times with a plastic comb. Bring the comb near your hair without touching it. Move it around.

2. **Experiment** Comb your hair again. Place the comb near a faucet of running water.

Drawing Conclusions

3. **Communicate** Describe your observations in steps 1 and 2.
_____
_____

4. Rub an inflated balloon against your hair. Place the balloon against a wall. What happens? Why?
_____
_____

Unit F · Energy   Use with textbook page F71   191

Name_____ **Date**_____

QUICK LAB
FOR SCHOOL OR HOME
Lesson 6

5. Going Further Think about other things to test about static electricity. Write and conduct an experiment.

My Hypothesis Is:

_____

_____

My Experiment Is:

_____

_____

_____

_____

_____

_____

My Results Are:

_____

_____

_____

_____

192   Unit F · Energy   Use with textbook page F71

Name_____ Date_____

What Makes a Bulb Light?

Hypothesize What parts are needed to make a light bulb light? How should they be arranged?

Write a **Hypothesis:**

_____

_____

Materials

- flashlight bulb
- 20 cm of wire with stripped ends
- 2 D-cells
- cell holder

Procedure

BE CAREFUL! Objects that produce light also produce heat.

1. **Experiment** Work with your group to try to light the bulb using the materials. Draw each setup. Record your results. Use another piece of paper for drawings.

   _____

2. **Predict** Study the drawings on this page. Predict in which setups the bulb will light and in which it will not light. Record your predictions.

   _____

   _____

3. **Experiment** Work with another group of students to test each setup. Can you see a pattern?

   _____

   _____

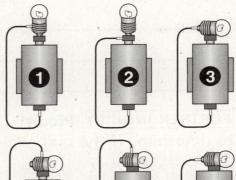
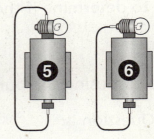
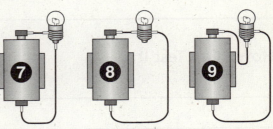
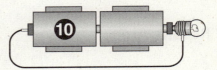

4. Place your hand close to each setup to see if the wires or bulbs produce heat. Be careful not to touch the wires or bulbs directly.

Unit F · Energy   Use with textbook page F77   **193**

Name_____ Date_____

Drawing Conclusions

1. **Observe** How many ways could you arrange the materials to make the bulb light in step 1?
   _____

2. In which drawings did the bulb light? How are the setups similar?
   _____
   _____

3. Which setups produced the most heat? Why do you think it is important to use electricity safely?
   _____
   _____
   _____

4. FURTHER INQUIRY **Predict** Draw another setup. Challenge a classmate to determine if the bulb will light. Use a separate piece of paper.

Inquiry

Do a test using the bulb, wire, and D-cell to make the setup fail.

My Question Is:
_____

How I Can Test It:
_____
_____

My Results Are:
_____
_____

194   Unit F · Energy   Use with textbook page F77

Name_____ Date_____

Complete Circuits

Procedure

Materials
- flashlight bulb
- 20 cm of wire with stripped ends
- D-cell
- cell holder

1. Your teacher will assign your group two of the circuit patterns shown in Lesson 7 Explore Activity. Draw the two patterns that were assigned to your group.

2. Assemble your first pattern. Did the bulb light?
   _____

3. Assemble your second pattern. Did the bulb light?
   _____

4. Share your results with the class. In the class results, look for similarities in the circuits that caused the bulb to light.

Drawing Conclusions

1. What did the circuits that made the bulb light have in common?
   _____
   _____

2. What did the circuits that didn't light the bulb have in common?
   _____
   _____

Unit F · Energy   Use with TE textbook page F77

Name_____ Date_____

Conductor Test-Off

Hypothesize The base and wires of a light bulb are good conductors. The filament is a poor conductor. What other materials are good conductors or insulators?

Write a **Hypothesis:**

_____
_____

Materials
- flashlight bulb
- bulb socket
- D-cell
- cell holder
- 3 wires with stripped ends, 20 cm each
- assorted test objects

Procedure

1. **Experiment** Make a circuit as shown, using a test object. Record your observations.

   _____
   _____

2. Test other objects. Record your observations.

   _____
   _____
   _____

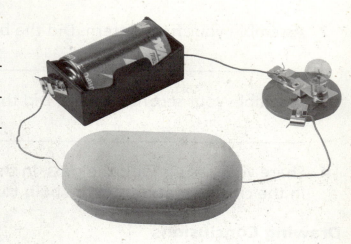

Drawing Conclusions

3. **Observe** Which objects were good conductors? Which were not? How could you tell?

   _____
   _____
   _____
   _____

Unit F · Energy   Use with textbook page F80

Name_____ **Date**_____

QUICK LAB
FOR SCHOOL OR HOME
Lesson 7

4. **Infer** Examine a length of wire. Which part of the wire is a conductor? Which part is an insulator? Why do you think the wire is made this way?

_____

_____

_____

5. **Going Further** If a second wire is connected to the circuit described in the Procedure, would the lightbulb light up? Write and conduct an experiment.

My Hypothesis Is:

_____

_____

My Experiment Is:

_____

_____

_____

My Results Are:

_____

_____

_____

_____

©Macmillan/McGraw-Hill

Unit F · Energy   Use with textbook page F80   **197**

Name_____ Date_____

Explore Activity
Lesson 8

How Is a Bar Magnet Like a Compass?

Hypothesize How does a bar magnet compare with a compass? How could you find out?

Write a **Hypothesis**:
_____
_____

Materials
- 2 bar magnets
- 1 m of string
- compass
- ruler
- tape
- heavy book

Procedure

1. **Observe** How do the bar magnets interact when you place them next to each other in different positions?
   _____
   _____

2. **Predict** Which way will the bar magnet point if you hang it as shown? Record your prediction.
   _____

3. **Observe** Test your prediction. Record the result.
   _____

4. Place the compass on a flat surface away from the magnets. Compare the directions in which the compass and magnet point.
   _____

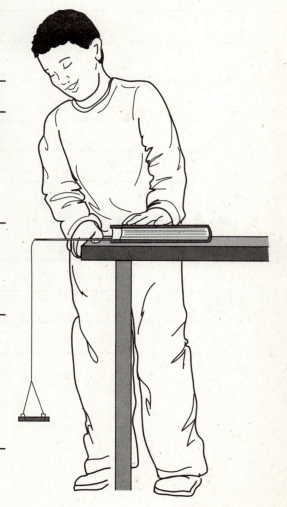

198   Unit F · Energy   Use with textbook page F89

Name_____ Date_____

Explore Activity
Lesson 8

5. **Observe** Hold the compass near the hanging magnet. What happens?
   _____

Drawing Conclusions

1. **Communicate** How do the two magnets interact with each other?
   _____
   _____

2. **Communicate** What happened when you brought the compass near the hanging magnet?
   _____
   _____

3. Of what must a compass be made?
   _____

4. **FURTHER INQUIRY** **Infer** Compare your hanging magnet to a compass. How are they alike? Different?
   _____
   _____

Inquiry

Think of your own questions that you might like to test using the bar magnets. What items will the magnets attract?

My Question Is:
_____

How I Can Test It:
_____

My Results Are:
_____

Unit F · Energy   Use with textbook page F89   199

Name_____ Date_____

Alternative Explore
Lesson 8

Pointing North

Procedure

BE CAREFUL! Be careful when handling the steel needle.

1. Hold the magnet near the sandwich bag. Do not unseal the bag! Observe what happens to the iron filings.

2. Stroke the needle several times in the same direction along the magnet.

3. Hold the needle near the sandwich bag. Observe what happens to the iron filings.

4. Fill the cup with water to within an inch of the top. Float the cork on the water. It should not touch the sides of the cup.

5. Place the needle on the cork. Be sure that the cork and needle are free to turn.

6. Compare the position of the needle with the position of the pointer in the compass.

_____

_____

Materials

- bar magnet
- steel needle
- piece of cork
- plastic drinking cup
- water
- compass
- iron filings in a sealed plastic sandwich bag

Drawing Conclusions

1. What did stroking the needle along the magnet do to the needle? How do you know?

_____

_____

2. How did the positions of the needle and the compass pointer compare?

_____

200   Unit F · Energy   Use with TE textbook page F89

Name _____ **Date** _____

QUICK LAB
FOR SCHOOL OR HOME
Lesson 8

Electromagnets

Hypothesize What will make an electromagnet stronger?
Write a **Hypothesis:**

_____

_____

Materials

- nail
- 2 D-cells and holders
- wire with stripped ends
- 10 paper clips

Procedure

1. Wind a wire 20 times around a nail near its head. Attach each end of the wire to a D-cell.

2. **Predict** How many paper clips do you think your electromagnet can hold? Test your prediction.

   _____

3. **Experiment** Repeat using two D-cells in series.

   _____

   _____

4. **Experiment** Repeat steps 1–3 winding the wire 20 more times.

   _____

   _____

©Macmillan/McGraw-Hill

Unit F · Energy

Use with textbook page F92

201

Name	Date

QUICK LAB
FOR SCHOOL OR HOME
Lesson 8

Drawing Conclusions

5. **Interpret Data** How did increasing current affect the strength of the electromagnet? Increasing the number of coils?

_____

_____

6. **Going Further** Which change has the greater effect—increasing current or adding coils? Explain. Write and conduct an experiment.

My Hypothesis Is:

_____

_____

My Experiment Is:

_____

_____

My Results Are:

_____

_____

202   Unit F · Energy   Use with textbook page F92

Name_____ Date_____

Inquiry Skill Builder
Lesson 8

Communicate

Send Messages with Electromagnetism

In this activity, you will use a buzzer to produce sounds. Electricity activates an electromagnet inside the buzzer to produce a buzzing sound.

Materials

- Mini buzzer with two leads
- 2 dry cells
- dry cell holder

Procedure

1. Insert two dry cells into their holders. Wrap the tip of the buzzer's black wire to the clip on the negative end of one of the cell holders.

2. **Experiment** Touch the tip of the red wire to the positive end of the second dry cell. Does it make a sound? Experiment making long and short sounds.

3. **Communicate** Have a partner make up questions. Use the chart to communicate your answers with the buzzer. Create more responses as you need them.

Sound on Buzzer	Meaning in Words
1 short beep	Yes
2 short beeps	No
1 short beep, 2 long beeps	I don't know

©Macmillan/McGraw-Hill

Unit F · Energy   Use with textbook page F93   203

| Name_____ | Date_____ |

Inquiry Skill Builder

Lesson 8

Drawing Conclusions

1. What did you do to make the buzzer sound? How did you make long beeps? Short beeps?

_____

_____

_____

2. **Infer** What would happen if the magnet, wire, or dry cell were removed from the electric circuit?

_____

_____

_____

3. **Communicate** How is a system of sounds that represents words a useful means of communication?

_____

_____

_____

204   Unit F · Energy   Use with textbook page F93